Environmental and Resources Geochemistry
of Earth System

Naotatsu Shikazono

Environmental and Resources Geochemistry of Earth System

Mass Transfer Mechanism, Geochemical Cycle and the Influence of Human Activity

Naotatsu Shikazono (deceased)
Keio University
Tokyo, Japan

Dr. James Wilkinson, colleague and longtime friend of the late Prof. Shikazono, proofread this book and compiled the list of symbols.

CHIKYUU SISUTEMU NO KAGAKU
CHIKYUU SISUTEMU KANKYOU KAGAKU
© 1997 Naotatsu SHIKAZONO
© 2010 Naotatsu SHIKAZONO
All rights reserved
Original Japanese edition published in 1997 and 2010 by University of Tokyo Press
English translation rights arranged with University of Tokyo Press through Japan UNI Agency, Inc., Tokyo

ISBN 978-4-431-54903-1   ISBN 978-4-431-54904-8 (eBook)
DOI 10.1007/978-4-431-54904-8
Springer Tokyo Heidelberg New York Dordrecht London

Library of Congress Control Number: 2014936099

© Springer Japan 2015
This work is subject to copyright. All rights are reserved by the Publisher, whether the whole or part of the material is concerned, specifically the rights of translation, reprinting, reuse of illustrations, recitation, broadcasting, reproduction on microfilms or in any other physical way, and transmission or information storage and retrieval, electronic adaptation, computer software, or by similar or dissimilar methodology now known or hereafter developed. Exempted from this legal reservation are brief excerpts in connection with reviews or scholarly analysis or material supplied specifically for the purpose of being entered and executed on a computer system, for exclusive use by the purchaser of the work. Duplication of this publication or parts thereof is permitted only under the provisions of the Copyright Law of the Publisher's location, in its current version, and permission for use must always be obtained from Springer. Permissions for use may be obtained through RightsLink at the Copyright Clearance Center. Violations are liable to prosecution under the respective Copyright Law.
The use of general descriptive names, registered names, trademarks, service marks, etc. in this publication does not imply, even in the absence of a specific statement, that such names are exempt from the relevant protective laws and regulations and therefore free for general use.
While the advice and information in this book are believed to be true and accurate at the date of publication, neither the authors nor the editors nor the publisher can accept any legal responsibility for any errors or omissions that may be made. The publisher makes no warranty, express or implied, with respect to the material contained herein.

Printed on acid-free paper

Springer is part of Springer Science+Business Media (www.springer.com)

Foreword

This book 'Environmental and Resource Geochemistry of Earth System' should be in the libraries of universities offering courses in mining engineering and/or mining geology and should be read by mine managers and directors and concerned politicians.

Chapter 1 provides an extensive discussion of chemical equilibria as they pertain to minerals. This covers in detail chemical equilibria of solids and aqueous solutions, the solubility of various types of minerals, the weathering of silicate and aluminosilicate minerals, the composition of geothermal and ground water, hydrothermal alteration, oxidation and reduction reactions, and the partitioning of elements between aqueous solutions and crystals.

Chapter 2 covers the partial chemical equilibria as they pertain to water–rock interaction in geothermal areas. This includes oxygen isotope variations, formation of minerals by boiling, due to mixing of fluids and due to mixing hydrothermal solutions with seawater or groundwater.

Chapter 3 provides a detailed discussion of mass transfer mechanisms.

Chapter 4 presents detailed discussions of hydrothermal and seawater systems, emphasizing the pertinent recharge, reservoir, and discharge zones, as well as the pertinent chemical reactions.

Chapter 5 provides a detailed discussion of the geochemical cycles of carbon, sulfur, and oxygen, as well as of arsenic, boron, barium, and other ore elements.

Chapter 6 discusses in great detail the interaction between humans and the atmosphere, hydrosphere, and soils, as well as the effects of human waste emissions.

It was a great honor to be invited to write the Foreword to this book by the late Prof. Naotatsu Shikazono, who sadly did not live to see the completion of this publication. Prof. Shikazono made extremely valuable contributions to his field and will be greatly missed by his friends and colleagues.

Ulrich Petersen, Ph.D.
Department of Earth and Planetary Sciences
Harvard University, Cambridge
MA, USA

Preface

The Earth system consists of the atmosphere, hydrosphere, lithosphere (geosphere), biosphere, and humans, and each subsystem interacts with other bodies with regard to mass and heat (Fig. 1). The interactions between humans and the other subsystems are becoming the most important problems for humans. The interaction between humans and nature poses difficulties because current methodologies do not yet offer solutions to its problems. For instance, people today are facing issues related to Earth's resources and environmental problems (e.g., depletion of resources; pollution of the atmosphere, hydrosphere, and soils; extinction of biomass; global warming; acid rain; and destruction of the ozone layer).

To resolve the environmental and resources problems, a scientific understanding of the Earth system is the first step. Particularly, mechanisms of mass and energy transport and circulation among the subsystems (water, atmosphere, rocks, i.e., minerals) should be scientifically understood. Among the various processes, the water–rock interactions near the Earth's surface significantly influence the formations of Earth's resources, the generation of environmental pollution, and elemental and mass transport near the Earth's surface environment.

In this book we consider (1) water–soils–rock reaction, (2) waste–groundwater reaction, (3) formation of metallic ore deposits, (4) mass transfer in the hydrothermal system, and (5) the global geochemical cycle. To interpret these processes, thermodynamics, dissolution-precipitation kinetics, fluid flow, diffusion and coupling of these mechanisms, mass transfer, and geochemical modeling (e.g., dissolution-kinetics fluid flow modeling) are used. In addition to dissolution, precipitation reactions, ion exchange reaction, adsorption, and biomineralization are briefly described. The analyses of global geochemical cycles of carbon, sulfur, and other elements are based on the box model.

This book has arisen from several courses on Environmental and Resources Geochemistry of Earth and Planetary System Sciences for undergraduate and graduate students at Keio University and also from many classes at other universities (The University of Tokyo, Akita University, Chiba University, Ibaraki University, Hiroshima University, Yamaguchi University, Nihon University, Tokushima University, Yamagata University, Kyoto University, Shizuoka University, Tsukuba

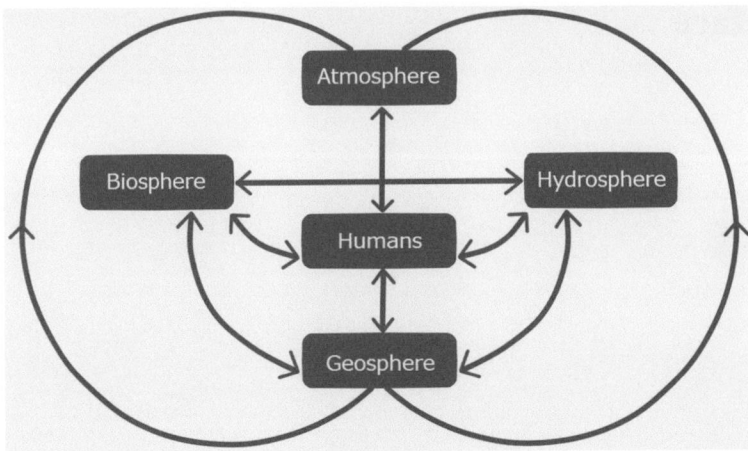

Fig. 1 Earth system consisting of subsystems such as atmosphere, biosphere, hydrosphere, humans and geosphere. Subsystems are interacting each other with regard to mass and heats

University, Tohoku University, Waseda University, Gakushuin University, Tokyo Gakugei University). In these courses during the last 30 years, I received many comments, questions, and responses from numerous undergraduate and graduate students. Discussions with them have allowed me to develop and clarify the ideas presented here. In writing this book I am greatly indebted to many people in the Department of Applied Chemistry of Keio University, the Geology Department of The University of Tokyo, the Geology Department of Tokyo Gakugei University, and the Department of Earth and Planetary Science of Harvard University. I express my great appreciation for the late professors T. Tatsumi, of the University of Tokyo, T. Fuji of Tsukuba University, A. Tsusue of Kumamoto University, and H. D. Holland and U. Petersen of Harvard University for teaching me economic geology, environmental geochemistry, thermodynamics and kinetics. Dr. K. Fujimoto and Dr. N. Takeno read the manuscript and gave me useful critical comments. I thank Ms. M. Aizawa, Ms. N. Katayama, Ms. K. Suga, Ms. N. Takeuchi, Ms. A. Takeuchi, and Mr. Y. Suga for their skillful and patient word processing and figure drawing. Ms. M. Shimizu and Ms. M. Komatsu of The University of Tokyo Press and Mr. Ken Kimlicka of Springer Japan edited with care the manuscripts for the Japanese version of the books *Chemistry of Earth System* and *Environmental Chemistry of Earth System* and the English version of *Environmental and Resources Geochemistry of Earth System* (the present volume), respectively.

I want to dedicate this book to my wife, Midori Shikazono, and my two daughters, Chikako and Hisako Shikazono; and to my parents, Yoshiko and Naoharu Shikazono, who have patiently provided understanding and moral support during the more than 45 years of my academic research and teaching.

Tokyo, Japan   Naotatsu Shikazono

Contents

1	**Chemical Equilibrium**	1
1.1	Chemical Equilibrium Model	1
1.2	Thermochemical Stability of Solid Phase in Solid-Aqueous Solution System	3
1.3	Solubility of Minerals	8
	1.3.1 Oxides and Hydroxides	8
	1.3.2 Silicates	8
	1.3.3 Carbonates	13
1.4	Chemical Weathering and Silicate and Aluminosilicate Mineralogy	15
1.5	Analysis of Multi-Component–Multi-Phase Heterogeneous System	17
1.6	Interpretation of Chemical Compositions of Geothermal Water in Terms of Chemical Equilibrium Model	18
1.7	Hydrothermal Alteration	23
1.8	Interpretation of Chemical Composition of Ground Water in Terms of Inverse Mass Balance Model	25
1.9	Oxidation-Reduction Condition	33
	1.9.1 Oxygen Fugacity (f_{O_2})-pH Diagram	33
	1.9.2 H–S–O System	33
	1.9.3 Fe–S–O–H System	36
	1.9.4 Estimate of Oxygen Fugacity (f_{O_2}) for Hydrothermal Ore Deposits	36
	1.9.5 Sulfides	38
1.10	Partitioning of Elements Between Aqueous Solution and Crystal	44
	1.10.1 Partitioning of Element Between Aqueous Solution and Solid Solution Mineral	44
	1.10.2 Rayleigh Fractionation	48
	Cited Literature	49

2 Partial Chemical Equilibrium ... 53
  2.1 Water–Rock Interaction ... 53
  2.2 Hydrothermal Alteration Process in Active and Fossil Geothermal Areas ... 55
  2.3 Oxygen Isotopic Variations During Water–Rock Reaction, Mixing and Boiling of Fluids ... 57
  2.4 Formation of Minerals Accompanied by Separation of Vapor Phase and Liquid Phase and Boiling ... 58
    2.4.1 One Step Boiling ... 60
    2.4.2 Multi-Step Boiling ... 60
  2.5 Precipitation of Minerals Due to Mixing of Fluids ... 61
  2.6 Formation of Hydrothermal Ore Deposits by Hydrothermal Solution-Seawater Mixing ... 65
    2.6.1 Mid-Oceanic Ridge Deposits ... 65
    2.6.2 Kuroko Deposits ... 66
  2.7 Formation of Gold Deposits by Mixing of Hydrothermal Solution and Acid Ground Water ... 67
  Cited Literature ... 70

3 Mass Transfer Mechanism ... 73
  3.1 Dissolution-Precipitation Kinetics ... 73
    3.1.1 Dissolution Mechanism ... 73
    3.1.2 Precipitation Mechanism ... 78
    3.1.3 Metastable Phase ... 82
  3.2 Diffusion ... 83
    3.2.1 Fick's Law ... 83
    3.2.2 Diffusion in Pore in Rocks and Minerals ... 83
  3.3 Advection ... 84
    3.3.1 Darcy's Law ... 84
    3.3.2 Three Dimensional Fluid Flow ... 86
  3.4 Coupled Models ... 87
    3.4.1 Reaction-Fluid Flow Model ... 87
    3.4.2 Reaction-Diffusion Model ... 95
    3.4.3 Diffusion-Flow Model ... 96
    3.4.4 Temperature-Dependent Model ... 99
  Cited Literature ... 99

4 System Analysis ... 103
  4.1 Hydrothermal System ... 103
    4.1.1 Recharge Zone ... 105
    4.1.2 Reservoir ... 107
    4.1.3 Discharge Zone ... 108
    4.1.4 Formation of Chimney and Sea-Floor Hydrothermal Ore Deposits and Precipitation Kinetics-Fluid Flow Model ... 110

Contents xi

		4.1.5	Precipitation-Dispersion Model	113
		4.1.6	Diffusion-Fluid Flow Model	114
		4.1.7	Dissolution-Recrystallization Model	116
		4.1.8	Formation of Metastable Phase	117
	4.2	Seawater System		118
		4.2.1	Chemical Equilibrium and Steady State	118
		4.2.2	Chemical Equilibrium Model	119
		4.2.3	Ion Exchange Equilibrium	121
		4.2.4	Factors Controlling Chemical Composition of Seawater (Input and Output Fluxes)	122
	Cited Literature			137
5	**Geochemical Cycle**			141
	5.1	General Equation		141
	5.2	Carbon Cycle		143
		5.2.1	Short-Term Cycle (Biogeochemical Cycle)	144
		5.2.2	Long-Term Cycle (Geochemical Cycle)	147
	5.3	Sulfur Cycle		151
		5.3.1	Short-Term Cycle	152
		5.3.2	Long-Term Cycle	153
	5.4	Coupled Geochemical Cycle: Sulfur–Carbon–Oxygen (S–C–O) Cycle		154
	5.5	Global Geochemical Cycle-Mass Transfer Between Earth's Surface System and Interior System		157
		5.5.1	Global Carbon Cycle	158
		5.5.2	Global Sulfur (S) Cycle	163
	5.6	Geochemical Cycle of Minor Elements		165
		5.6.1	Arsenic (As)	165
		5.6.2	Boron (B)	167
		5.6.3	Barium (Ba)	168
		5.6.4	Other Ore Constituent Elements	168
	Cited Literature			170
6	**Interaction Between Nature and Humans**			173
	6.1	Flux to the Atmosphere Due to Human Activity		173
		6.1.1	Carbon Dioxide (CO_2)	173
		6.1.2	Sulfur (S)	174
		6.1.3	Phosphorus (P)	176
		6.1.4	Minor Elements	177
		6.1.5	Geochemical Cycles of Pb, Cd and Hg Have Been Well Investigated Because of Their High Toxicity	179
	6.2	Anthropogenic Fluxes to the Hydrosphere and Soils and Mass Transfer Mechanism		182
		6.2.1	Acid Rain-Soil-Ground Water System	182
		6.2.2	Pollution of River Water	193

		6.2.3	Pollution of Lake Water	194
		6.2.4	Pollution in Ocean	200
	6.3	Feedback Associated with Human Waste Emissions		201
		6.3.1	Geological Disposal of High Level Nuclear Waste	201
		6.3.2	Underground CO_2 Sequestration	208
	Cited Literature			210

Afterwords 215

Appendix (Plate) 219

Index 239

Introduction

The earth system is divided into the fluid earth and the solid earth. The fluid earth consists of subsystems such as the atmosphere and hydrosphere, and the solid earth consists of subsystems such as the crust, mantle and core. The chemical and physical characteristics of each subsystem (composition, volume, mass, chemical state etc.) have been extensively investigated and, consequently, are described only briefly in the text below.

The atmosphere that envelopes the earth consists primarily of gases (nitrogen, N_2 78 %; oxygen, O_2 21 %; argon, Ar 0.93 %). Minor components include water vapor, H_2O (40–40,000 ppm by volume), CO_2 (390 ppm), neon, Ne (18.2 ppm), and helium, He (5.24 ppm).

The hydrosphere consists mainly of oceanic and terrestrial freshwater in the liquid phase. Compared to terrestrial water environments, such as rivers, lakes and ground water, the chemical composition of the oceans is relatively constant. The amount of water contained in the soil, in minerals and rocks (fluid inclusions, adsorbed surface water, water in crystals etc.), water component of organisms, and water vapor in the atmosphere, all of which are found near the earth's surface environment, is minor.

The solid earth consists of rocks composed of minerals and small amounts of rock in the amorphous phase (e.g., volcanic glass). The solid earth has three components; the crust, the mantle and the core. The outer core is molten and the inner core is solid in state (iron and nickel). The solid (rocks) forms the mantle and crust.

The crust extends from the surface of the earth to the Mohorovicic's (Moho) discontinuity, which is the first plane of unconformity, or boundary, between the mantle and crust. The crust is divided into the oceanic crust and the continental crust. The oceanic crust is mainly composed of basic and ultrabasic rocks (e.g., basalt, gabbro, and peridotite) overlain by marine sediments. Continental crust is more siliceous (av. SiO_2: 66 %) than the oceanic crust (av. SiO_2: 50 %) and the average chemical composition of the oceanic crust is roughly equivalent to basalt: granite = 1:1. It is also composed of various types of rocks (igneous rocks, metamorphic rocks, sedimentary rocks).

The mantle extends from Mohorovicic's discontinuity to a depth of 2,900 km. It is divided into the upper and lower mantles based on seismic velocity measurements. The mantle consists of silicates (e.g., olivine, garnet, perovskite), and it has a high MgO content.

The core of the earth lies more than approximately 2,890 km below the surface and is divided into an inner core and an outer core. S waves do not travel through the outer core, indicating that the outer core is in the liquid phase. The inner core is solid and composed predominantly of iron and nickel.

The biosphere, humans, surface water, soils and atmosphere all exist near the boundary between the fluid earth and the solid earth (crust). In this book, this zone is referred to as the earth's surface environment. These subsystems interact with each other and their characteristic features change over time.

The biosphere consists of the sum total of all organisms on earth and is minuscule (0.25×10^{-9}) compared to the mass of the earth (Holland and Petersen 1995). However, the biosphere has a significant influence upon the surface environment of the earth (i.e. atmosphere, hydrosphere, soils).

This book focuses on the interactions between the hydrosphere (mainly water) and the geosphere (mainly rocks) in the earth's surface environment, i.e., water-rock interactions, such as weathering, hydrothermal alteration and mineralization (formation of ore deposits and associated elemental migration).

Chemical equilibrium and mass transfer mechanisms (chemical reactions, diffusion, fluid flow (advection), adsorption, etc.) (Chaps. 1, 2, and 3) are examined in order to illustrate the compositional variation that exists within water (ground water, hydrothermal solution, seawater) and weathered and hydrothermally altered rocks and soils. To better understand the subsystems of the earth, equilibrium and mass transfer coupling models are applied to the seawater system, as an example of a low-temperature exogenic system, and hydrothermal systems, as an example of high-temperature endogenic systems (Chap. 4).

A box model is presented that considers the geochemical cycle and attempts to explain the earth system (Chap. 5). Although such models are typically not used to consider mass transfer mechanisms, this chapter examines how box models can be employed to improve our understanding of global geochemical cycles. For example, previously published geochemical cycle models (e.g., BLAG model by Berner et al. 1983) consider the CO_2 cycling in the atmosphere-hydrosphere-crust system, and exclude the mantle (Chap. 5). However, in this book, the mantle is also considered, as it is an important reservoir in global geochemical cycles of the earth system. In addition, degassing via subduction zones is an important component of the global geochemical cycle and climate change (Sect. 5.2.2).

Interactions between fluid and solid phases occur in low-temperature systems (e.g., weathering, ground water, seawater) and high-temperature systems (e.g., hydrothermal system). Previously, biogeochemical cycle dynamics have generally only been considered within the context of low temperatures. This is the first book on earth systems to consider both low and high temperature cycle between low and high temperature cycle dynamics.

Further, although humanity constitutes only a small part of the biosphere, human activities have a marked influence the surface environment of the earth (atmosphere, hydrosphere, biosphere, soils). Consequently, the influence of humans on the other subsystems (atmosphere, hydrosphere, soils, rocks) through, for example, acid rain, pollution of surface water, geological disposal of nuclear waste, and underground CO_2 sequestration are also examined (Chap. 6).

Cited Literature

Berner RA, Lasaga AC, Garrels RM (1983) Am J Sci 283:641–683
Holland HD, Petersen U (1995) Living dangerously. Princeton University Press, Princeton

Further Reading

Ernst WG (ed) (2000) Earth systems: processes and issues. Cambridge University Press, Cambridge
Holland HD, Petersen U (1995) Living dangerously. Princeton University Press, Princeton
Jacobson MC, Charlson RJ, Rodhe H, Orians GH (2000) Earth system science from biogeochemical cycles to global change. Elsevier, Amsterdam
Mason B (1958) Principles of geochemistry, 2nd edn. Wiley, New York
Shikazono N (2012) Introduction to earth and planetary system science. Springer, Berlin

Symbols and Variables Used in This Text

a, α	Activity
A	Surface area
C	Heat capacity
D	Diffusion coefficient
E	Yang module, activation energy
F	Flux, formation factor
f	Fugacity, constant, fanning factor
G	Gibbs free energy
g	Gravity
I.A.P	Ionic activity product
I	Ionic strength
J	Heat flux
k	Doener–Hoskins distribution coefficient, rate constant, mass transfer coefficient, precipitation rate constant, hydraulic conductivity, reaction rate constant, heat conductivity, dissolution rate constant
k′, k″	Rate constant of dissolution
Kd	Distribution coefficient, partition coefficient
L	Mole ratio
m	Molality, cation concentration, concentration, solubility
N	Avogadro number
ø	Porosity, potential
P	Partial pressure, pressure
q	Flow rate
Q	Reaction activity quotient
r	Ionic radius
R	Precipitation rate, dissolution rate, gas constant, resistivity (Ohms)
R	Rock, retardation coefficient
Re	Reynolds number
S.I.	Saturation index
S	Seawater

T	Absolute temperature
t	Time
u	Mass of dissolved solid
V	Volume
v	Volume, velocity
w	Sedimentation rate
W	Weight
z	Charge
β	Specific-ion parameters
γ	Activity coefficient
δ	A small change in the variable, width of a fracture
κ	Equilibrium constant, solubility product, permeability
λ	Activity coefficient
μ	Chemical potential, Poisson ratio, viscosity coefficient, viscosity
ρ	Density
ς	Progress variable
τ	Tortuosity, time
υ	Stoichiometric coefficient, mol coefficient, kinematic viscosity coefficient, velocity, volume
χ	Concentration (Mole Fraction)
Ψ	Fluid potential
Ω	Degree of super-saturation, saturation index, saturation quotient

Chapter 1
Chemical Equilibrium

1.1 Chemical Equilibrium Model

Natural system is composed of gas, aqueous liquid (water) and solid phases. In order to understand the natural system chemical equilibrium model is the most useful and has been applied to this system. The principles of chemical equilibrium model and examples of the application of this model will be described below.

Figure 1.1 shows chemical equilibrium model of the natural system. Variables which determine the thermochemical feature of this system include temperature, total pressure, activities of dissolved species in aqueous solution (ions, ion pairs, complexes etc.), gaseous fugacity, activities of components in solid phases and dissolved species in aqueous solution where activity of i species, a_i, is equal to $\gamma_i m_i$ (m_i is molality of i species in aqueous solution and mole fraction of i component in solid solution and γ_i is activity coefficient of i species in aqueous solution and of each component in solid phase).

Assuming chemical equilibrium between aqueous solution and solid phase at constant temperature and pressure, concentration of dissolved species could be estimated. In order to estimate the concentration the values of other variables (concentration and activity coefficient of each component in solid phase and activity coefficient of dissolved species in aqueous solution) have to be estimated.

The concentration of each component in solid phase is determined by chemical analysis of solid phase. Activity coefficient of each component in solid phase is deduced from the relation, $a_i = \gamma_i X_i$ where a_i is activity of i component, X_i is concentration (mole fraction) of i component, and γ_i is activity coefficient of i component.

If above relation can not be determined, it is sometimes possible to deduce activity coefficient values for components in solid phase based on thermodynamic solid solution model. For example, if symmetrical solvus (Fig. 1.2) exists for a binary system, regular solution model could be applicable to the estimation of activity coefficients and other thermodynamic parameters values of solid solution

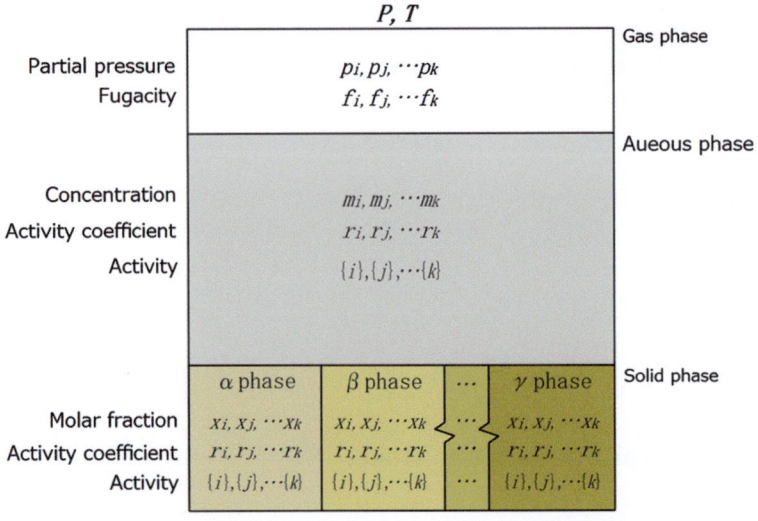

Fig. 1.1 Generalized thermochemical equilibrium model for natural system (Stumm and Morgan 1970)

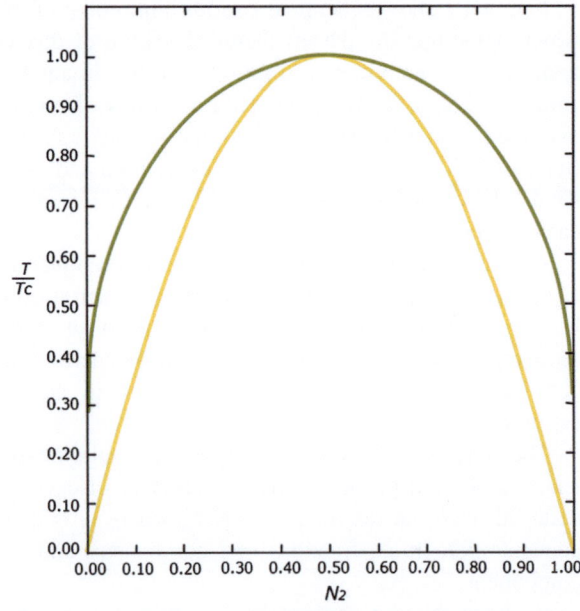

Fig. 1.2 Solvus (*solid line*) and spinodal (*dashed*) for a strictly regular solution ($W_s = 0$), assuming W_H to be constant. Phases within the spinodal are unstable with respect to internal diffusion $[(\partial^2 G/\partial N_2^2)_{p11} < 0]$ (Thompson 1967)

(e.g., Guggenheim 1952, 1967; Thompson 1967). While the regular solid solution model predicts symmetric miscibility gaps, the actual gaps are frequently asymmetric like for strontianite-aragonite (Sr, Ca) CO_3 solid solution (e.g., Glynn and Reardon 1990).

1.2 Thermochemical Stability of Solid Phase in Solid-Aqueous Solution System

Debye–Hückel theory predicts activity coefficient (γ_i) of ionic species in low salinity aqueous solution from the equation that is expressed as

$$-\log\gamma_i = Az_i^2 I^{1/2}/\left(1 + aBI^{1/2}\right) \quad (1.1)$$

where z_i is charge of i species, I is ionic strength ($I = 1/2\sum z_i^2 m_i$ where m is molal concentration), A and B are constant for a given solvent at constant temperature and pressure and a is constant for each ion related to ionic radii. The Debye–Hückel theory can be applied to the dilute aqueous solution with ionic strength less than 0.2.

Several extended Debye–Hückel equations have been proposed for saline aqueous solution (Stumm and Morgan 1981). For example, Davies equation is expressed as follows (Davies 1962)

$$\log\gamma_i = -Az_i^2\left\{I^{1/2}/\left(1 + I^{1/2}\right) - 0.31\right\} \quad (1.2)$$

Pitzer's equation (the Virial equation) which can be applied to higher salinity solution than Davies equation is given by (Pitzer 1991)

$$\ln\gamma_{\pm} = (-A/3)|z^+ z^-|f(I) + (2v^+v^-/v)B(I)m + 2\left\{(v^+v^-)^{3/2}/v\right\}Cm^2 \quad (1.3)$$

where $f(I) = I^{1/2}(1 + 1.2I^{1/2}) + 1.671n(1 + 1.2I^{1/2})$, and $B(I) = 2\beta^o + (2\beta'/\alpha^2 I)(1 - (1 + \alpha I^{1/2} - (1/2)\alpha^{21})e^{-\alpha I^{1/2}}$

where m is the molality, z is the charge of the designated species, v is stoichiometric coefficient in the normal formula of the MX salt, β^o, and β' are specific-ion parameters, α is a constant value for similarly charged class of electrolytes, α is a specific-ion parameter independent of ionic strength, and B and α is second and third virial coefficient, respectively, The parameter, $v = v^+ v^-$ is the sum of the stoichiometric coefficients for the cation, v^+, and the anion, v^- of the solute, z is charge, and m is molal concentration.

Pitzer's formulation allows the calculation of the thermodynamic properties of concentrated aqueous solutions, up to solubility limits (e.g., Harvie et al. 1984).

1.2 Thermochemical Stability of Solid Phase in Solid-Aqueous Solution System

Thermochemical stability fields of solid phases in simple system are considered below.

At first, the stability field of K_2O (solid phase) in K_2O–H_2O system is derived. The chemical reaction between K_2O and aqueous solution is represented by

$$K_2O + 2H^+ = H_2O + 2K^+ \tag{1.4}$$

Equilibrium constant for (1.4) is given by

$$K_{1\text{-}4} = (a_{H_2O}a_{K^{+2}})/(a_{K_2O}a_{H^{+2}}) \tag{1.5}$$

where a is activity.

Assuming a_{H_2O} and a_{K_2O} is unity, we obtain

$$\log(a_{K^+}/a_{H^+}) = 1/2 \log K_{1\text{-}4} \tag{1.6}$$

Therefore, (1.6) and (1.7) indicate that the stability field of K_2O is represented as a function of a_{K^+}/a_{H^+} and $K_{1\text{-}4}$ that is represented as a function of temperature and pressure. The a_{K_2O} relates to chemical potential of K_2O (μ_{K_2O}) as follows.

$$\mu_{K_2O} = \mu_{0K_2O} + RT \ln a_{K_2O} \tag{1.7}$$

where μ_{0K_2O} is standard chemical potential of K_2O.

Therefore, the stability field of K_2O is also represented by μ_{K_2O}, μ_{K^+}, and μ_{H^+}.

Generally, natural minerals are not oxides such as K_2O. Instead, potassium is contained in silicates such as K-feldspar ($KAlSi_3O_8$) and K-mica ($KAl_3Si_3O_{10}(OH)_2$). In addition to potassium, sodium is present in silicates like feldspar and mica. Al_2O_3, SiO_2, and H_2O are usually contained in silicates in addition to K_2O and Na_2O.

Next, we consider the thermochemical stability fields of minerals in K_2O–Na_2O–SiO_2–Al_2O_3–H_2O system. The stabilities of K-mica and Na-montmorillonite are represented by the equilibrium relation for the following reaction.

$$2.3 KAl_3Si_3O_{10}(OH)_2 \text{(K-mica)} + Na^+ + 1.3H^+ + 4SiO_2 \text{(quartz)}$$
$$= 3Na_{0.33}Al_{2.33}Si_{3.66}O_{10}(OH)_2 \text{(Na-montmorillonite)} + 2.3K^+ \tag{1.8}$$

Equilibrium constant for reaction (1.8) is given by

$$K_{1\text{-}8} = a_{K^+}^{2.3}/a_{Na^+} a_{H^+}^{1.3} \tag{1.9}$$

where activities of solid phases are assumed to be unity.

Using these equations, the stability fields of the minerals in this system can be shown on activity diagram and chemical potential diagram (Figs. 1.3, 1.4, and 1.5). Thermochemical data that can be used for constructing stability diagrams are presented in Table 1.1.

Fig. 1.3 Activity diagram for Na_2O–K_2O–Al_2O_3–H_2O system; 250 °C, 1Kb (Henley 1984). Solid circle represents chemical composition of typical geothermal water

Fig. 1.4 Schematic stability relations in chemical potential (μ) diagram for K_2O–Na_2O–Al_2O_3–SiO_2–H_2O–HCl system at 400 °C and 1 Kb. Pyrophyllite is metastable phase (Rose and Burt 1979)

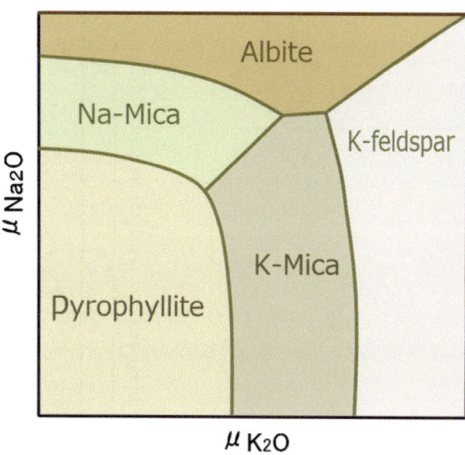

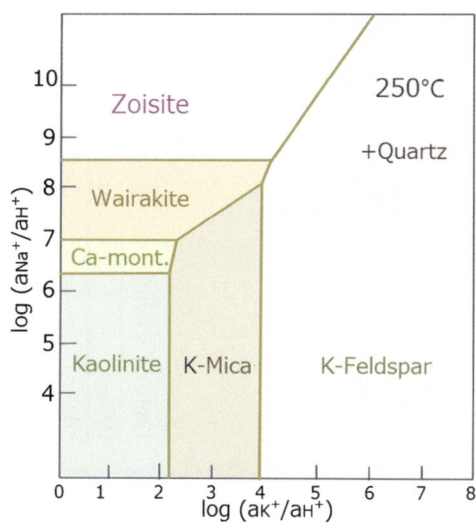

Fig. 1.5 Activity diagram for the principal phases in the CaO–Al_2O_3–K_2O–H_2O system at 250 °C (Henley 1984)

Table 1.1 Equilibrium constant for geothermal water—mineral reaction (Arnórsson et al. 1982)

Mineral	Reaction	Equilibrium constant (K)
Adularia	$KAlSi_3O_8 + 8H_2O = K^+ + Al(OH)_4^- + 3H_4SiO_4$	$+38.85 - 0.0548T - 172,620/T + 1,012,722/T^2$
Low temperature Na-feldspar	$NaAlSi_3O_8 + 8H_2O = Na^+ + Al(OH)_4^- + 3H_4SiO_4$	$+36.83 - 0.0439T - 16,474/T + 1,004,631/T^2$
Anhydrite	$CaSO_4 = Ca^{2+} + SO_4^{2-}$	$+6.20 - 0.0229T - 1,217/T$
Calcite	$CaCO_3 = Ca^{2+} + CO_3^{2-}$	$+10.22 - 0.0349T - 2,476/T$
Chalcedony	$SiO_2 + 2H_2O = H_4SiO_4$	$+0.11 - 1,101/T$
Mg-chlorite	$Mg_5Al_2Si_3O_{10}(OH)_8 + 10H_2O = 5Mg^{2+} + 2Al(OH)_4^- + 3H_4SiO_4 + 8OH^-$	$-1,022.12 - 0.3861T + 412.46\log T$
Fluorite	$CaF_2 = Ca^{2+} + 2F^-$	$+66.54 - 4,318/T - 25.47 \log T$
Goethite	$FeOOH + H_2O + OH^- = Fe(OH)_4^-$	$+80.34 - 0.099T + 20,290/T - 2,179,296/T^2$
Laumontite	$CaAlSi_4O_{12} \cdot 4H_2O + 8H_2O = Ca^{2+} + 2Al(OH)_4^- + 4H_4SiO_4$	$+65.95 - 0.0828T - 28,358/T + 1,916,098/T^2$
Microcline	$KAlSi_3O_8 + 8H_2O = K^+ + Al(OH)_4^- + 3H_4SiO_4$	$+44.55 - 0.0498T - 19,883/T + 1,214,019/T^2$
Magnetite	$Fe_3O_4 + 4H_2O = 2Fe(OH)_4^- + Fe^{2+}$	$-155.58 + 0.1658T + 35,298/T + 4,258,774/T^2$
Ca-montmorillonite	$6Ca_{0.167}Al_{2.33}Si_{3.67}O_{10}(OH)_2 + 60H_2O + 12OH^- = Ca^{2+} + 14Al(OH)_4^- + 22H_4SiO_4$	$+30,499.49 + 3.5109T - 1,954,295/T + 125,536,640/T^2 - 10,715.66\log T$
K-montmorillonite	$3K_{0.167}Al_{2.33}Si_{3.67}O_{10}(OH)_2 + 30H_2O + 6OH^- = K^+ + 7Al(OH)_4^- + 11H_4SiO_4$	$+15,075.11 + 1.7346T - 967,127/T + 61,985,927/T^2 - 5,294.72\log T$

1.2 Thermochemical Stability of Solid Phase in Solid-Aqueous Solution System

Mineral	Reaction	log K
Mg-montmorillonite	$6Mg_{0.167}Al_{2.33}Si_{3.67}O_{10}(OH)_2 + 60H_2O + 12OH^- = Mg^{2+} + 14Al(OH)_4^- + 22H_4SiO_4$	$+30,514.87 + 3.5188T - 1,953,843/T + 125,538,830/T^2 - 10,723.71\log T$
Na-montmorillonite	$3Na_{0.33}Al_{2.33}Si_{3.67}O_{10}(OH)_2 + 30H_2O + 6OH^- = Na^+ + 7Al(OH)_4^- + 11H_4SiO_4$	$+15,273.90 + 1.7623T - 978,782/T + 62,805,036/T^2 - 5,366.18\log T$
Mica	$KAl_3Si_3O_{10}(OH)_2 + 10H_2O + 2OH^- = K^+ + 3Al(OH)_4^- + 3H_4SiO_4$	$+6,113.68 + 0.6914T - 394,755/T + 25,226,323/T^2 - 2,144.77\log T$
Pyrrhotite	$8FeS + SO_4^{2-} + 22H_2O + 6OH^- = 8Fe(OH)_4^- + 9H_2S$	$+3,014.68 + 1.2522T - 103,450/T - 1,284.86\log T$
Pyrite	$8FeS + 26H_2O + 10OH^- = 8Fe(OH)_4^- + SO_4^{2-} + 15H_2S$	$+4,523.89 + 1.6002T - 180,405/T - 1,860.33\log T$
Quartz	$SiO_2 + 2H_2O = H_4SiO_4$	$+0.41 - 1.309/T$ (0–250 °C); $+0.12 - 1,164/T$ (180–300 °C)
Wairakite	$CaAl_2Si_4O_{12} \cdot 2H_2O + 10H_2O = Ca^{2+} + 2Al(OH)_4^- + 4H_4SiO_4$	$+61.00 - 0.0847T - 25,018/T + 1,801,911/T^2$
Wollastonite	$CaSiO_3 + 2H^+ + H_2O = Ca^{2+} + H_4SiO_4$	$-222.85 - 0.0337T + 16,258/T^2 + 80.68\log T$
Zoisite	$Ca_2Al_3Si_3O_{12}(OH) + 12H_2O = 2Ca^{2+} + 3Al(OH)_4^- + 3H_4SiO_4 + OH^-$	$+106.61 - 0.1497T - 40,448/T + 3,028,977/T^2$
Epidote	$Ca_2FeAl_2Si_3O_{12} + 12H_2O = 2Ca^{2+} + Fe(OH)_4^- + 2Al(OH)_4^- + 3H_4SiO_4 + OH^-$	$-27,399.84 - 3.8749T + 1,542,767/T - 91,778,364T^2 + 9,850.38\log T$
Marcasite	$8FeS_2 + 26H_2O + 10OH^- = 8Fe(OH)_4^- + SO_4^{2-} + 15H_2S$	$+4,467.61 + 1.5879T - 16,944/T - 1,838.45\log T$

1.3 Solubility of Minerals

Minerals dissolve into aqueous solution in a batch system and generally the concentrations increase with time and reach at constant value that is equilibrium concentration (solubility).

The solubility of minerals depends on chemical composition of aqueous solution, temperature and pressure. The dependence of solubility of minerals (oxides, hydroxides, silicates, carbonates and sulfides) on physicochemical variables (pH, P_{CO_2} etc.) is derived below.

1.3.1 Oxides and Hydroxides

The solubility of oxides and hydroxides strongly depends on pH at constant temperature and pressure.

The solubility of oxides and hydroxides can be estimated using equilibrium constants for the dissolution of hydroxides and oxides and chemical reactions in aqueous solutions listed in Table 1.2.

For example, ZnO dissolves into aqueous solution as Zn^{2+}, $ZnOH^+$, $Zn(OH)_{3-}$, and $Zn(OH)_4^{2-}$. The stability of each dissolved Zn species depends on pH at constant temperature and pressure. The solubility of ZnO can be derived from the chemical equilibria for the following chemical reactions.

$$\left. \begin{array}{l} ZnO + 2H^+ = Zn^{2+} + H_2O \\ ZnO + H^+ = ZnOH^+ \\ ZnO + 2H_2O = Zn(OH)_{3-} + H^+ \\ ZnO + 3H_2O = Zn(OH)_4^{2-} + 2H^+ \end{array} \right\} \quad (1.10)$$

The dependence of solubilities of ZnO, CuO and amorphous $Fe(OH)_3$ on pH can be calculated using thermochemical data given in Table 1.2 and is shown in Fig. 1.6.

Figure 1.6 indicates the hydroxides and oxides solubilities as influenced by hydrolysis reactions and they increase with a decrease in pH in acid region, and increase with an increase in pH in alkaline region.

1.3.2 Silicates

The solubility of silicates in CO_2— bearing solution as functions of pH and P_{CO_2} (partial pressure of CO_2 gas) is derived below (Stumm and Morgan 1970).

Na-feldspar ($NaAlSi_3O_8$) (Appendix, Plate 1) dissolves into CO_2— bearing solution, accompanied by the precipitation of kaolinite. This reaction is written as

1.3 Solubility of Minerals

Table 1.2 Constants for solubility equilibria of oxides, hydroxides, carbonates and hydroxide carbonates (Stumm and Morgan 1970)

Reaction	Log K (25 °C)	I
$H_2O(l) = H^+ + OH^-$	−14.00	0
	−13.77	1 M NaClO$_4$
$(am)Fe(OH)_3(s) = Fe^{3+} + 3OH^-$	−38.7	3 M NaClO$_4$
$(am)Fe(OH)_3(s) = FeOH^{2+} + 2OH^-$	−27.5	3 M NaClO$_4$
$(am)Fe(OH)_3(s) = Fe(OH)_2^+ + OH^-$	−16.6	3 M NaClO$_4$
$(am)Fe(OH)_3(s) + OH^- = Fe(OH)_4^-$	−4.5	3 M NaClO$_4$
$2(am)Fe(OH)_3(s) = Fe_2(OH)_2^{4+} + 4OH^-$	−51.9	3 M NaClO$_4$
$(am)FeOOH(s) + 3H^+ = Fe^{3+} + 2H_2O$	3.55	3 M NaClO$_4$
$\alpha\text{-FeOOH}(s) + 3H^+ = Fe^{3+} + 2H_2O$	1.6	3 M NaClO$_4$
$\alpha\text{-Al(OH)}_3(\text{gibbsite}) + 3H^+ = Al^{3+} + 3H_2O$	8.2	0
$\gamma\text{-Al(OH)}_3(\text{bayerite}) + 3H^+ = Al^{3+} + 3H_2O$	9.0	0
$(am)Al(OH)_3(s) + 3H^+ = Al^{3+} + 3H_2O$	10.8	0
$Al^{3+} + 4OH^- = Al(OH)_4^-$	32.5	0
$Cu(s) + 2H^+ = Cu^{2+} + H_2O$	7.65	0
$Cu^{2+} + OH^- = CuOH^+$	6.0 (18 °C)	0
$2Cu^{2+} + 2OH^- = Cu_2(OH)_2^{2+}$	17.0 (18 °C)	0
$Cu^{2+} + 3OH^- = Cu(OH)_3^-$	15.2	0
$Cu^{2+} + 4OH^- = Cu(OH)_4^{2-}$	16.1	0
$ZnO(s) + 2H^+ = Zn^{2+} + H_2O$	11.18	0
$Zn^{2+} + OH^- = ZnOH^+$	5.04	0
$Zn^{2+} + 3OH^- = Zn(OH)_3^-$	13.9	0
$Zn^{2+} + 4OH^- = Zn(OH)_4^{2-}$	15.1	0
$Cd(OH)_2(s) + 2H^+ = Cd^{2+} + 2H_2O$	13.61	0
$Cd^{2+} + OH^- = CdOH^+$	3.8	1 M LiClO$_4$
$Mn(OH)_2(s) = Mn^{2+} + 2OH^-$	−12.8	0
$Mn(OH)_2(s) + OH^- = Mn(OH)_3^-$	−5.0	0
$Fe(OH)_2(\text{active}) = Fe^{2+} + 2OH^-$	−14.0	0
$Fe(OH)_2(\text{inactive}) = Fe^{2+} + 2OH^-$	−14.5 (−15.1)	0
$Fe(OH)_2(\text{inactive}) + OH^- = Fe(OH)_3^-$	−5.5	0

I ionic strength

$$NaAlSi_3O_8(\text{Na-feldspar}) + H^+ + 9/2H_2O$$
$$= Na^+ + 2H_4SiO_4 + 1/2Al_2Si_2O_5(OH)_4(\text{kaolinite}) \quad (1.11)$$

Logarithm of equilibrium constant for this reaction is given by

$$\log K_{1\text{-}11} = \log(a_{Na} + a_{H_4SiO_4^2})/a_{H^+} = -1.9 \quad (1.12)$$

where activity of solid phase (Na-feldspar, kaolinite) and liquid H_2O is assumed to be unity.

CO_2(gas) dissolves into H_2O (liquid) as

$$CO_2 + H_2O = HCO_3^- + H^+ \quad (1.13)$$

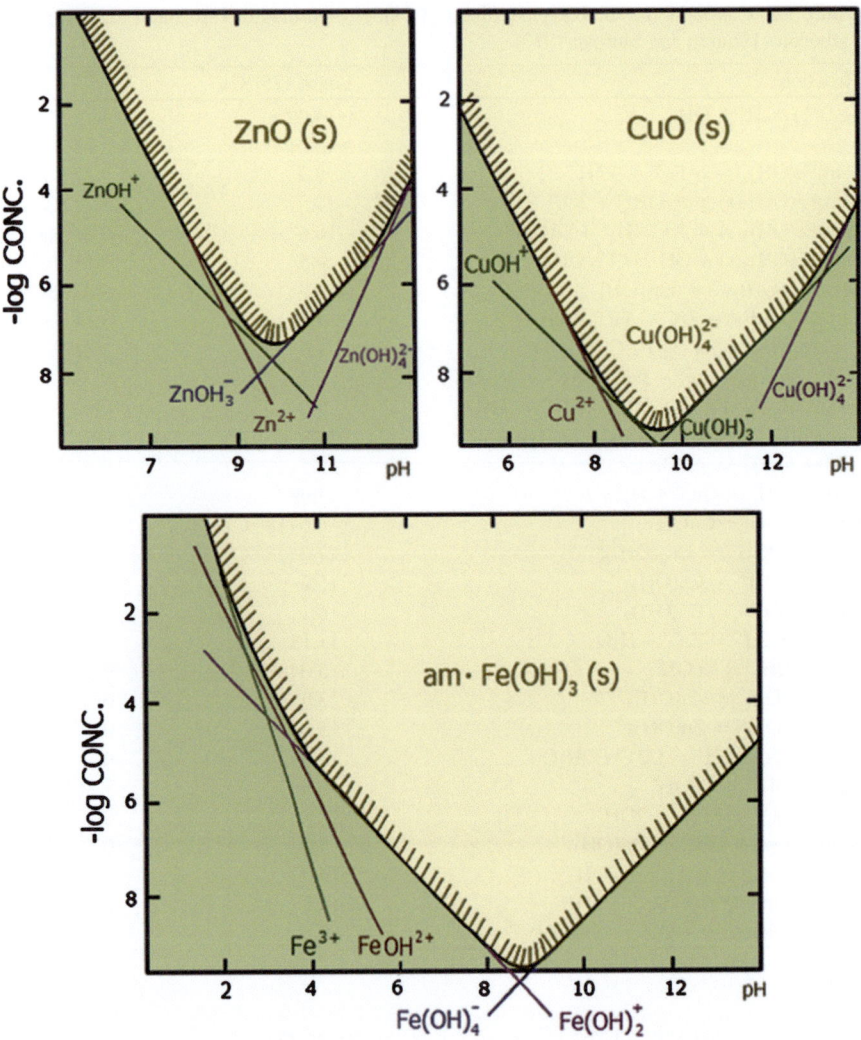

Fig. 1.6 Solubility of amorphous Fe(OH)$_3$, ZnO and CuO. The possible occurrence of polynuclear complexes, for example, Fe$_2$(OH)$_2^{4+}$, Cu$_2$(OH)$_2^{2+}$, has been ignored. Such complexes do not change the solubility characteristics markedly for the solids considered here (Stumm and Morgan 1970)

Logarithm of equilibrium constant for this reaction is

$$\log K_{1\text{-}13} = \log(a_{\text{HCO}_3^-}\, a_{\text{H}^+})/P_{\text{CO}_2} = -7.8 \tag{1.14}$$

Combining (1.11) with (1.13), we obtain

$$\begin{aligned} \text{NaAlSi}_3\text{O}_8 + \text{CO}_2 + 11/2\text{H}_2\text{O} \\ = \text{Na}^+ + \text{HCO}_3- + 2\text{H}_4\text{SiO}_4 + 1/2\text{Al}_2\text{Si}_2\text{O}_5(\text{OH})_4 \end{aligned} \tag{1.15}$$

1.3 Solubility of Minerals

Equilibrium constant for this reaction is given by

$$\log K_{1\text{-}15} = \log\left(a_{\text{Na}^+} a_{\text{HCO}_3^-} a_{\text{H}_4\text{SiO}_4^2}\right)/P_{\text{CO}_2} = -9.7 \quad (1.16)$$

Assuming Na^+ and HCO_3^- are dominant cation and anion species, respectively, electroneutrality relation is approximately expressed as

$$m_{\text{Na}^+} = m_{\text{HCO}_3^-} \quad (1.17)$$

Stoichiometric relation for the congruent dissolution of Na-feldspar is given by

$$m_{\text{Na}^+} = 1/2 \, m_{\text{H}_4\text{SiO}_4} \quad (1.18)$$

From above equations, we obtain

$$m_{\text{HCO}_3^-} = P_{\text{CO}_2} \left(10^{-9.7}\right)^{1/4} = 3.8 \times 10^{-3} P_{\text{CO}_2} \quad (1.19)$$

Equation (1.19) indicates that the solubility of Na-feldspar (m_{Na^+}, $m_{\text{HCO}_3^-}$) is expressed as a function of P_{CO_2} and pH (Fig. 1.7).

The relationships between the solubilities of other silicates, and P_{CO_2} and pH can be derived by a similar manner. The decreasing order of solubilities is; Ca-feldspar ($\text{CaAl}_2\text{Si}_2\text{O}_8$) < Na-feldspar($\text{NaAlSi}_3\text{O}_8$) < Na-montmorillonite ($\text{Na}_{0.33}\text{Al}_{2.33}\text{Si}_{3.66}\text{O}_{10}(\text{OH})_2$) < quartz($\text{SiO}_2$) (Appendix, Plate 2), K-mica($\text{KAl}_3(\text{Si}_3\text{O}_{10})_2$) < gibbsite ($\gamma$-$\text{Al}(\text{OH})_3$), kaolinite($\text{Al}_2\text{Si}_2\text{O}_5(\text{OH})_4$) < hematite($\text{Fe}_2\text{O}_3$) (Fig. 1.7). It is noteworthy that this order is consistent with Goldich-Jackson's weathering series which represents the order of the stability of weathering minerals based on the geological observation of weathered rocks (Goldich 1938) (Fig. 1.8).

Goldich-Jackson's weathering series and solubility calculations indicate that mobilities of Ca^{2+} and Na^+ during chemical weathering are relatively large, but those of Al and Fe are small, and those of Mg, Mn, and Si are intermediate. In the natural environment stoichiometric relation is not established because many solid phases are present and the assumption, $m_{\text{Na}^+} = m_{\text{HCO}_3^-}$, is not established. In fact, ion pairs and complexes in aqueous solution have to be taken into account in addition to free cations and anions. Therefore, the solubility of silicates in the system consisting of multi-components and multi-phases has to be considered.

In the case of the calculation of K-feldspar solubility, it is assumed that $m_{\text{K}^+} = m_{\text{HCO}_3^-}$ according to the method by Stumm and Morgan (1970). However, in general, Na^+ is predominant among cations. Therefore, it can be approximated as

$$m_{\text{Na}^+} = m_{\text{HCO}_3^-} \quad (1.20)$$

If the chemical equilibrium for the following ion exchange reaction is attained

$$\text{Na-mineral} + \text{K}^+ = \text{K-mineral} + \text{Na}^+ \quad (1.21)$$

the following relation is obtained, assuming $a_{\text{Na-mineral}}$ and $a_{\text{K-mineral}}$ are unity

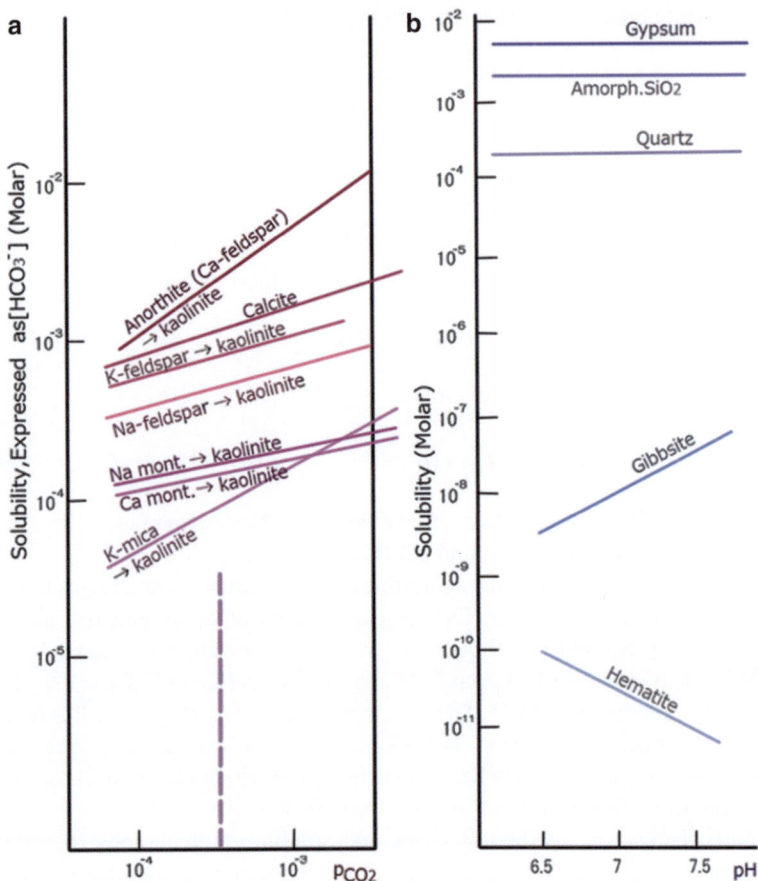

Fig. 1.7 Solubility of minerals (Stumm and Morgan 1970). (**a**) P_{CO_2} dependent solubility of "pure" minerals. Equilibrium [HCO_3^-] in reactions such as albite(s) + CO_2(g) + 11/2H_2O = Na^+ + HCO_3^- + 2H_4SiO_4 + $^1/_2$kaolinite(s) or calcite(s) + CO_2(g) + H_2O = Ca^{2+} + 2HCO_3^-; is used to express tendency for dissolution, (**b**) Congruent solubility of some minerals in the neutral pH range

$$K_{1\text{-}21} = a_{Na^+}/a_{K^+} \qquad (1.22)$$

Inserting (1.20) into (1.22), we obtain

$$m_{K^+} = (m_{HCO_3^-}\gamma_{Na^+})/(K_{1\text{-}21}\gamma_{K^+}) \qquad (1.23)$$

If Na^+ is predominant among aqueous cation species, then the solubility of K-mineral is expressed as Eq. (1.23).

The solubilities of the other minerals can be also derived based on the similar manner as mentioned above.

1.3 Solubility of Minerals

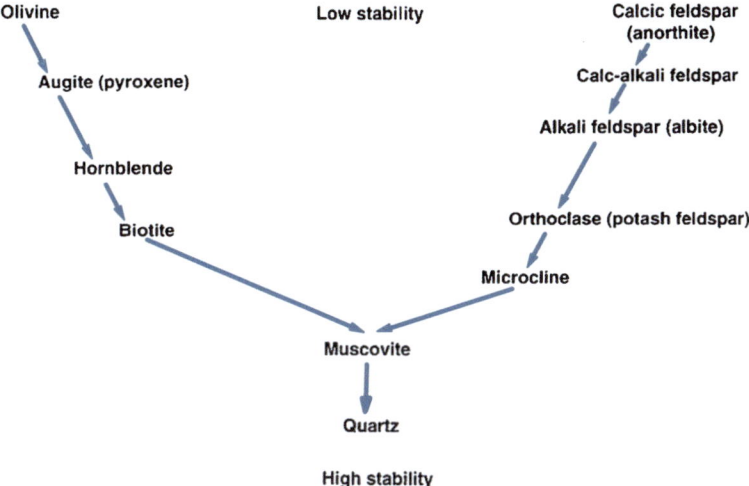

Fig. 1.8 Goldich stability series of mineral in the earth's surface (Lasaga 1998)

1.3.3 Carbonates

1.3.3.1 Solubility of Calcite at Constant P_{CO_2} (Partial Pressure of CO_2)

The solubility of calcite ($CaCO_3$) (Appendix, Plate 3) in H_2O at atmospheric P_{CO_2} is determined by the following reactions.

$$H_2CO_3 = H^+ + HCO_3^- \tag{1.24}$$

$$HCO_3^- = H^+ + CO_3^{2-} \tag{1.25}$$

$$CO_2 + H_2O = H_2CO_3 \tag{1.26}$$

$$CaCO_3 = Ca^{2+} + CO_3^{2-} \tag{1.27}$$

$$H_2O = H^+ + OH^- \tag{1.28}$$

Equilibrium constants for above reactions are given by

$$K_{1\text{-}24} = m_{H^+} m_{HCO_3^-} / m_{H_2CO_3} \tag{1.29}$$

$$K_{1\text{-}25} = m_{H^+} m_{CO_3^{2-}} / m_{HCO_3^-} \tag{1.30}$$

$$K_{1\text{-}26} = m_{H_2CO_3} / P_{CO_2} \tag{1.31}$$

$$K_{1\text{-}27} = m_{Ca^{2+}} m_{CO_3^{2-}} \tag{1.32}$$

$$K_{1\text{-}28} = m_{H^+} m_{OH^-} \tag{1.33}$$

where activity coefficients are assumed to be unity.

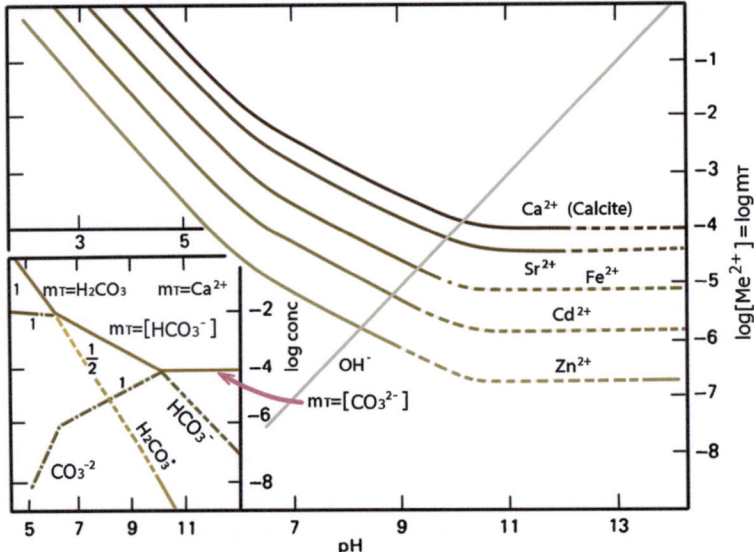

Fig. 1.9 Solubility of carbonates in a closed system: $[Me^{2+}] = C_T$ (total dissolved concentration of Me). The inset gives the essential features for the construction of the diagram for $CaCO_3(s)$ and equilibrium concentrations of all the carbonate species. A suspension of $MeCO_3(s)$ ($C_B - C_A = 0$) is characterized by the intersection of $[OH^-]$ and $[Me^{2+}] = C_T$. Dashed portions of the curves indicate conditions under which $MeCO_3(s)$ is not thermodynamically stable

Electroneutrality relation is represented by

$$2m_{Ca^{2+}} + m_{H^+} = m_{HCO_3^-} + 2m_{CO_3^{2-}} + m_{OH^-} \tag{1.34}$$

From (1.29), (1.30), (1.31), (1.32), (1.33), and (1.34) and giving $P_{CO_2} = 10^{-3.5}$ (atmospheric P_{CO_2}), we obtain pH = 8.3, $m_{H_2CO_3} = 10^{-5}$ m, $m_{HCO_3^-} = 10^{-3}$ m, $m_{Ca^{2+}} = 5 \times 10^{-4}$ m and $m_{CO_3^{2-}} = 1.6 \times 10^{-5}$.

1.3.3.2 Solubility of Calcite in Closed System

The concentration of total dissolved carbon species ($\sum C$) in aqueous solution in equilibrium with calcite in a closed system is constant which is expressed as

$$\left. \begin{array}{l} \sum C = m_{H_2CO_3} + m_{HCO_3^-} + m_{CO_3^{2-}} \\ m_{Ca^{2+}} = K_{SO}/m_{CO_3^{2-}} = K_{SO}/\sum C\alpha \end{array} \right\} \tag{1.35}$$

where $K_{SO} = m_{Ca^{2+}} m_{CO_3^{2-}}$, and α is proportion of CO_3^{2-} concentration to $\sum C$. The equilibrium saturation value of Ca^{2+} as a function of $\sum C$ and pH can be calculated (Fig. 1.9) (Stumm and Morgan 1996). Similar calculations can be done dor $MeCO_3$ (Me: Sr^{2+}, Fe^{2+}, Zn^{2+}, Mn^{2+} etc.; Mn CO_3, rhodochrosite (Appendix, Plate 4)). The graphical representation is given in a $\log[Me^{2+}]$ versus pH diagram (Fig. 1.9).

1.4 Chemical Weathering and Silicate and Aluminosilicate Mineralogy

Chemical weathering in a natural process that occurs under the prevailing environmental conditions, resulting in the transfer of matter from unstable minerals to more stable minerals and soluble species. Chemical weathering reactions include dissolution, precipitation, adsorption, and desorption of elements adsorbed on the mineral surfaces, ion exchange, oxidation, reduction and hydrolysis reactions.

Goldich (1938) examined the mineral assemblages present in soil (Appendix, Plate 5) under a variety of environmental conditions and established a stability series for sand and silt-sized particles that illustrates the relative stability of primary silicate minerals (Goldich's weathering series) (Fig. 1.8). For example, Ca-plagioclase, olivine (Appendix, Plate 6) and pyroxene (Appendix, Plate 7) tend to be most easily suffered chemical weathering and quartz and mica are most resistant to the weathering. This order is quite consistent with calculated solubility (Fig. 1.7) and experimentally determined dissolution rate of silicate minerals (Fig. 1.10). The solubility and dissolution rate of silicate minerals are related to the crystal structures, which is described below.

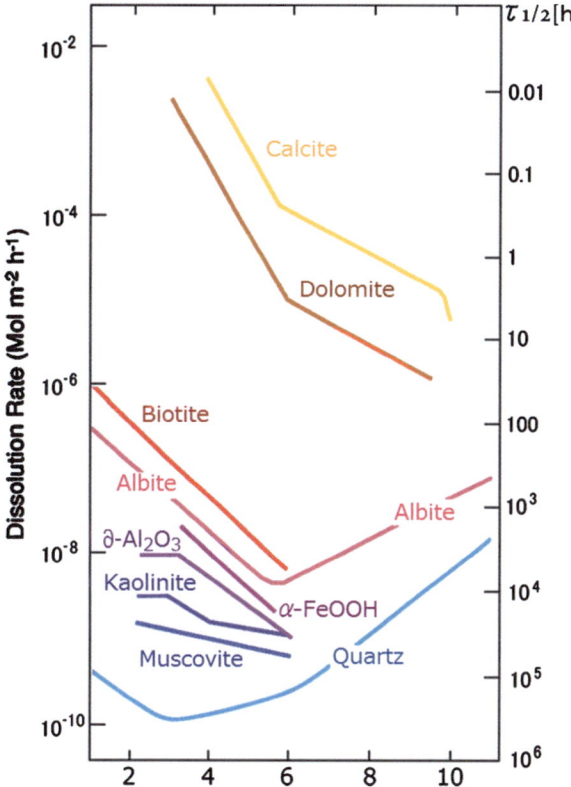

Fig. 1.10 Dissolution rate constant of minerals as a function of pH (Bidoglio and Stumm 1994)

Fig. 1.11 Si–O tetrahedron

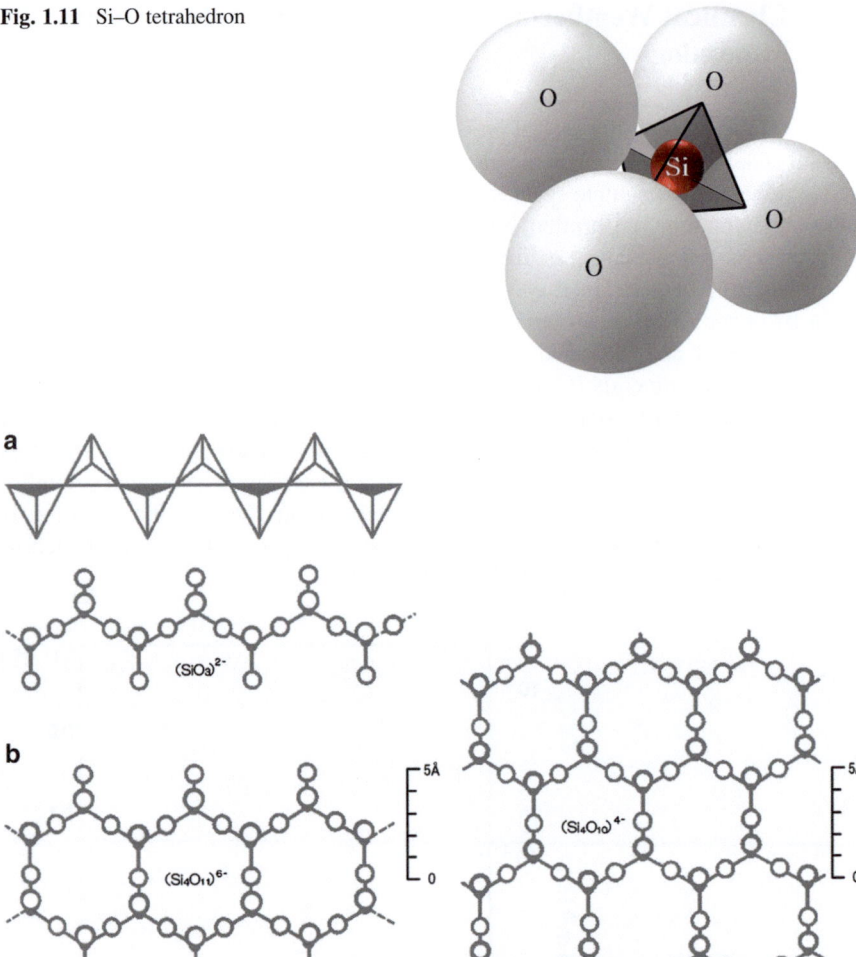

Fig. 1.12 Coordination model of a tetrahedrally polymerized sheet. Perfect cleavage parallel to the silicon–oxygen tetrahedral layers totally avoids the bridging oxygens. These planar structural units characterize the micas, chlorites and clay minerals. Weak bonding across the layers accounts for their softness and causes many sheet silicates to decompose readily at elevated temperatures. *A* angstroms (Ernst 1969)

Silicates and aluminosilicates are the most abundant minerals, which are main constituents of various rocks in the crust and mantle. Si^{4+} combines with O^{2-} to form silicate anion, SiO_4^{4-} in silicate crystal. The SiO_4^{4-} has tetrahedral form (Fig. 1.11). The SiO_4 tetrahedron is the basic building block of silicates in which silicon is situated at the center of a tetrahedron of four oxygen atoms. SiO_4^{4-} tetrahedra exist as discrete units or is joined via the O atoms, sharing their particles with other tetrahedra (Fig. 1.12). Silicate minerals mainly consisting of oxygen and

Table 1.3 The structural classification of the silicates (Mason 1952)

Classification	Structural arrangement	Silicon:oxygen ratio	Examples
Nesosilicates	Independent tetrahedral	1:4	Forsterite, Mg_2SiO_4
Sorosilicates	Two tetrahedral sharing one oxygen	2:7	Akermanite, $Ca_2MgSi_2O_7$
Cyclosilicates	Closed rings of tetrahedral each sharing two oxygens	1:3	Beryl, $Al_2Be_3Si_6O_{18}$
Inosilicates	Continuous single chains of tetrahedral each sharing two oxygens	1:3	Pyroxenes, e.g., enstatite, $MgSiO_3$
	Continuous double chains of tetrahedral sharing alternately two and three oxygens	4:11	Amphiboles, e.g., anthophyllite, $Mg_7(Si_4O_{11})_2(OH)_2$
Phyllosilicates	Continuous sheets of tetrahedral each sharing three oxygens	2:5	Talc, $Mg_3Si_4O_{10}(OH)_2$ Phlogopite, $KMg_3(AlSi_3O_{10})(OH)_2$
Tektosilicates	Continuous framework of tetrahedral each sharing all four oxygens	1:2	Quartz, SiO_2 Nepheline, $NaAlSiO_4$

silicon are classified into several groups according to the crystal structures. They include orthosilicates (no corners shared), disilicates (one corner shared), ring structure (two corners shared), single chains (two corners shared), double chains (two corners shared), sheet silicates (three corners shared), and framework silicates (four corners shared). Examples of each silicate type are shown in Table 1.3. Common silicate minerals include olivine, pyroxene and quartz.

Aluminum is the abundant element in the earth's crust. It occurs as cation, Al^{3+} and anion, $(AlO_4)^{5-}$ in minerals. The Al^{3+} is able to substitute for some of the Si^{4+} ions in silicate structure, forming aluminosilicates. Al^{3+} also substitutes for other cations such as Fe^{2+} and Mg^{2+}. Common aluminosilicate minerals include feldspars, micas, and amphiboles.

1.5 Analysis of Multi-Component–Multi-Phase Heterogeneous System

Natural system consists of multi-components and multi-phases. Major components of silicates include SiO_2, Al_2O_3, K_2O, CaO, MgO, FeO, Fe_2O_3, TiO_2, MnO and P_2O_5. Aqueous solution contains dissolved species such as ion pairs (NaCl, KCl, Na_2SO_4, K_2SO_4 etc.), cations (Na^+, K^+, Ca^{2+}, Fe^{2+}, Fe^{3+}, Mg^{2+} etc.), and anions (Cl^-, HCO_3^-, SO_4^{2-}, etc.). The concentrations of these species control the stability of minerals. In contrast, the mineral phases containing multi-components determine the compositions of aqueous solutions.

In order to analyze multi-component–multi-phase system, electroneutrality relation in aqueous solution has to be taken into account. This is given by

$$\sum m_i z_i = \sum m_j z_j \tag{1.36}$$

where m_i is concentration of cation i species, m_j is concentration of anion j species, and z_i and z_j is charge of cation i and anion j species, respectively.

Mass balance equation for a given element in aqueous solution is also used. For example, the total concentration of dissolved sulfur species ($\sum S$) is approximately given by

$$\sum S = m_{H_2S} + m_{HS^-} + m_{S^{2-}} + m_{SO_4^{2-}} + m_{HSO_4^-} \tag{1.37}$$

Ionic strength and thermochemical properties of solid solutions are necessary to be obtained to estimate activity coefficients of aqueous species and components in solid solutions.

The above equations (mass action law, extended Deye–Hückel equation, electroneutrality relation, mass balance equations) can be solved by several procedures such as graphic method, and Newton–Raphson method (Stumm and Morgan 1970).

Examples of interpretation of natural water chemistry in multi-component–multi-phase system are given below.

1.6 Interpretation of Chemical Compositions of Geothermal Water in Terms of Chemical Equilibrium Model

Geothermal system consists of recharge zone, reservoir (reaction zone), discharge zone and heat source (Fig. 1.13). Surface water (meteoric water, seawater) penetrates from recharge zone to reservoir. Water in reservoir is heated by heat source such as magma and hot rocks, resulting to the generations of hot water (hydrothermal solution) and vapor. Hot water and vapor ascend along cracks, fractures and fissures in rocks to ground surface.

Flow rate of hydrothermal solution is slow in geothermal reservoir. Chemical equilibrium between hydrothermal solution and surrounding rocks in geothermal reservoir is mostly attained after long time interaction between geothermal water and reservoir rocks. The water–rock interaction causes the changes in the chemical and mineralogical compositions of rocks (hydrothermal alteration) and chemical composition of geothermal water.

The chemical compositions of geothermal water are interpreted in terms of chemical equilibrium model between aqueous solution and altered rocks (Shikazono 1978a, b).

1.6 Interpretation of Chemical Compositions of Geothermal Water in Terms... 19

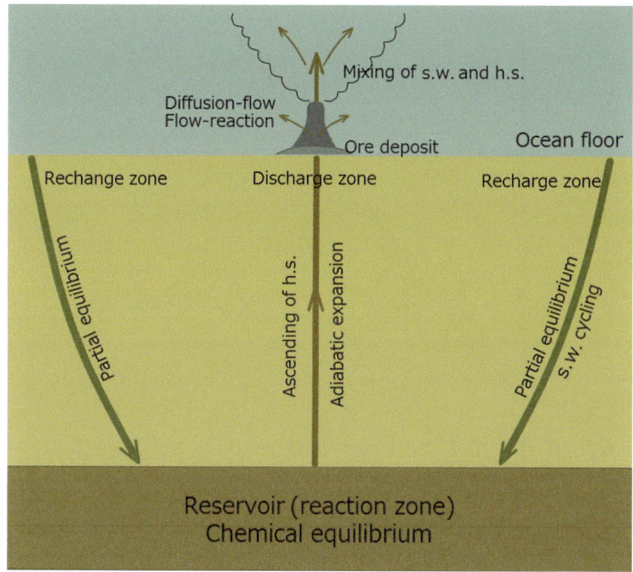

Fig. 1.13 Mass transfer process in submarine hydrothermal system (Shikazono 2010). *s.w.* seawater, *h.s.* hydrothermal solution

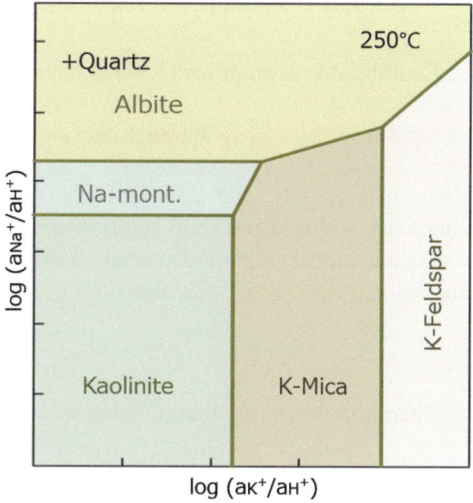

Fig. 1.14 Activity diagram for the principal phases in the Na_2O–K_2O–Al_2O_3–H_2O system at 250 °C (Henley 1984). *Solid circle* represents typical chemical composition (activity) of geothermal waters

Figure 1.14 shows typical chemical composition of geothermal water in equilibrium with dominant hydrothermal alteration minerals (quartz, Na-feldspar, K-feldspar, K-mica). Therefore, the chemical compositions of geothermal water in equilibrium with hydrothermal alteration minerals (quartz, Na-feldspar, K-feldspar, calcite, Ca-feldspar, anhydrite, Mg-chlorite) are derived below.

The chemical equilibrium among Na-feldspar, K-feldspar and geothermal water is derived using the reaction such as

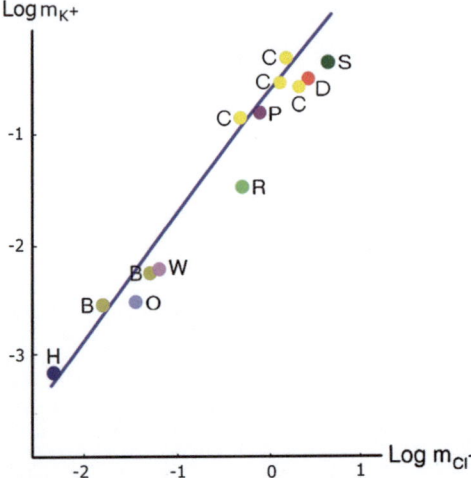

Fig. 1.15 Relation between the K$^+$ and Cl$^-$ concentration of geothermal waters and inclusion fluids. The *solid line* defines the equilibrium condition between the solution and the assemblage albite-K-feldspar at 250 °C (Shikazono 1978a, b). *Solid* and *open circles* mean the chemical analytical data on inclusion fluids and geothermal waters, respectively. *S* Salton sea, *R* Reykjanes, *W* Wairakei, *B* broadlands, *O* Otake, *H* Hveragerdi, *C* climax, *D* Darwin, *P* Providencia

$$NaAlSi_3O_8(\text{Na-feldspar}) + K^+ = KALSi_3O_8(\text{K-feldspar}) + Na^+ \qquad (1.38)$$

Equilibrium constant for (1.38) is expressed as

$$K_{1\text{-}28} = a_{Na^+}/a_{K^+} = m_{Na^+}/m_{K^+} \qquad (1.39)$$

where a is activity, and m is molality. Activities of Na-feldspar and K-feldspar components in feldspar solid solution are assumed to be unity. $\gamma_{Na^+}/\gamma_{K^+}$ (activity coefficient ratio) is regarded as unity. Assuming Na$^+$ and Cl$^-$ are predominant cation and anion, respectively, electroneutrality relation is approximately represented by

$$m_{Na^+} = m_{Cl^-} \qquad (1.40)$$

Therefore, from (1.39) and (1.40) we obtain

$$\log m_{K^+} = \log m_{Cl^-} - \log K_{1\text{-}28} \qquad (1.41)$$

This equation indicates that K$^+$ concentration (in logarithmic unit) increases with increasing Cl$^-$ concentration (in logarithmic unit) with a slope of +1 on $\log m_{K^+} + \log m_{Cl^-}$ diagram (Fig. 1.15).

The reaction among K-feldspar, quartz, K-mica and aqueous solution is represented by

1.6 Interpretation of Chemical Compositions of Geothermal Water in Terms...

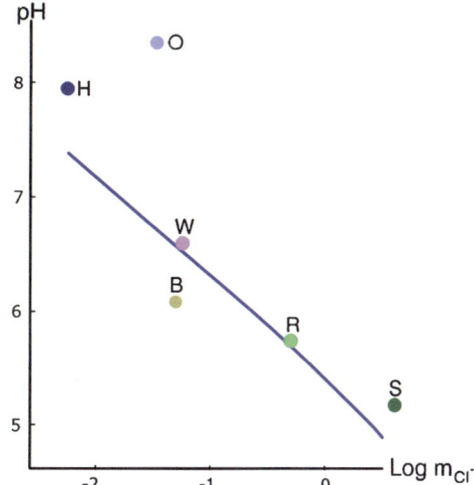

Fig. 1.16 Relation between the pH and Cl⁻ concentrations of geothermal waters. The *solid line* indicates the albite-K-feldspar-muscovite-quartz-solution equilibrium at 250 °C (Shikazono 1978a, b). For symbols used, see caption to Fig. 1.15

$$3KAlSi_3O_8(\text{K-feldspar}) + 2H^+$$
$$= KAL_3Si_3O_{10}(OH)_2(\text{K-mica}) + 6SiO_2(\text{quartz}) + 2K^+ \quad (1.42)$$

Using the equilibrium relation for (1.42), we obtain the relation between pH ($= -\log m_{H^+}$) and Cl⁻ concentration (in logarithmic unit) (Fig. 1.16). Figure 1.16 shows that pH decreases with increasing Cl⁻ concentration with a slope of approximately +1.

Similarly, we can derive the relationship between Ca^{2+} concentration and Cl⁻ concentration (Fig. 1.17). Above treatment indicates that chemical compositions of geothermal water are mostly consistent with the results of thermochemical calculations, indicating that the chemical equilibrium between geothermal water and alteration minerals is attained in geothermal reservoir (Shikazono 1978a, b). In the above discussion temperature and pressure of system are assumed to be constant. However, chemical composition of geothermal water ascending along the cracks deviates from the equilibrium between geothermal water and alteration minerals in the host rocks due to the change in temperature by the mixing of geothermal water and cold ground water and boiling of geothermal water (see Chap. 2).

Fournier and Truesdell (1973) obtained Na–K–Ca geothermometer representing the relationship between Na, K and Ca concentrations and temperature of geothermal water. Shikazono (1976) derived this relationship based on the chemical equilibrium among Ca–Na–K-feldspars, calcite and wairakite.

Silica geothermometer is also useful for an estimate of reservoir temperature because silica concentration of geothermal water which strongly depends on temperature due to the dependence of solubility of silica minerals (quartz, opal, chalcedony, amorphous silica) on temperature (Fig. 1.18). Temperature estimated

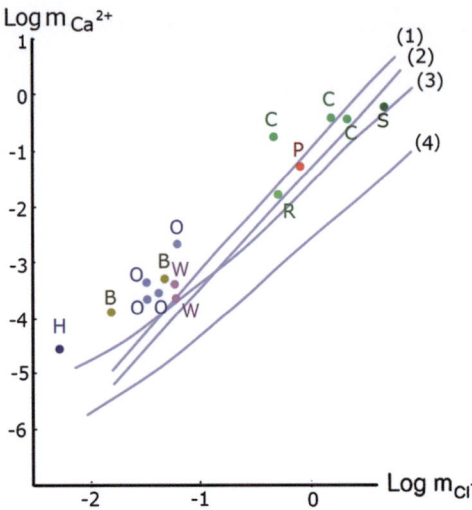

Fig. 1.17 Relation between the Ca^{2+} and Cl^- concentrations of geothermal waters and inclusion fluids. The *solid line* indicates: (*1*) albite-K-feldspar-muscovite-quartz-calcite-solution equilibrium at $a_{H_2CO_3} = 10^{-2.5}$; (*2*) albite-K-feldspar-muscovite-quartz-calcite-solution equilibrium at $a_{H_2CO_3} = 10^{-2}$; (*3*) anhydrite-solution at $\sum S_0$ (total dissolved sulfur concentration) $= 10^{-3}$; (*4*) anhydrite-solution equilibrium at $\sum S_0 = 10^{-2}$ (Shikazono 1978a, b). For symbols used, see caption to Fig. 1.15

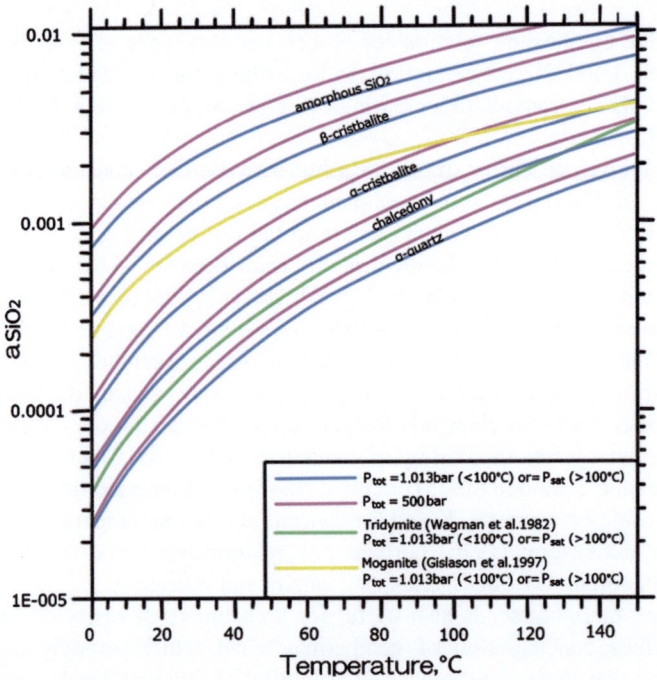

Fig. 1.18 Temperature dependence of the solubility of α-quartz, chalcedony, α-cristobalite, moganite, β-cristobalite and amorphous silica at different pressures, as specified (data computed by SUPCRT92). Also shown is the temperature dependence of the solubility of an unspecified tridymite (data from Wagman et al. 1982) (Marini 2007)

from silica geothermometer is generally in agreement with that from Na–K–Ca geothermometer, but frequently it deviates with each other probably due to the different dissolution rates of silica minerals and feldspars(see Chap. 3).

1.7 Hydrothermal Alteration

Hydrothermal alteration is the change in chemical and mineralogical compositions of rocks and in chemical compositions of hydrothermal solution during the water–rock interactions.

Hydrothermal alteration is classified into several types depending on the alteration minerals (Hemley and Jones 1964; Rose and Burt 1979). Important hydrothermal alterations include propylitic alteration, argillic alteration, advanced argillic alteration and sericitic alteration (Meyer and Hemley 1967).

Propylitic and advanced argillic alterations are commonly found in active geothermal systems and mining areas.

Propylitic alteration is characterized by Na-feldspar ($NaAlSi_3O_8$), K-feldspar ($KAlSi_3O_8$), epidote ($Ca_2FeAl_2Si_3O_{12}(OH)$), calcite ($CaCO_3$), prehnite ($Ca_2Al_2Si_3O_{10}(OH)_2$), chlorite (Mg-chlorite: $Mg_5Al_2Si_3O_{10}(OH)_8$, Fe-chlorite: $Fe_5Al_2Si_3O_{10}(OH)_8$ etc.), smectite (Ca-smectite: $Ca_{0.16}Al_{2.33}Si_{3.67}O_{10}(OH)_2$, Na-smectite: $Na_{0.33}Al_{2.33}Si_{3.66}O_{10}(OH)_2$, K-smectite: $K_{0.33}Al_{2.33}Si_{3.66}O_{10}(OH)_2$ etc.) which are formed in altered igneous rocks such as andesite. Propylitic alteration minerals change with depth and away from intrusive rocks (Fig. 1.19).

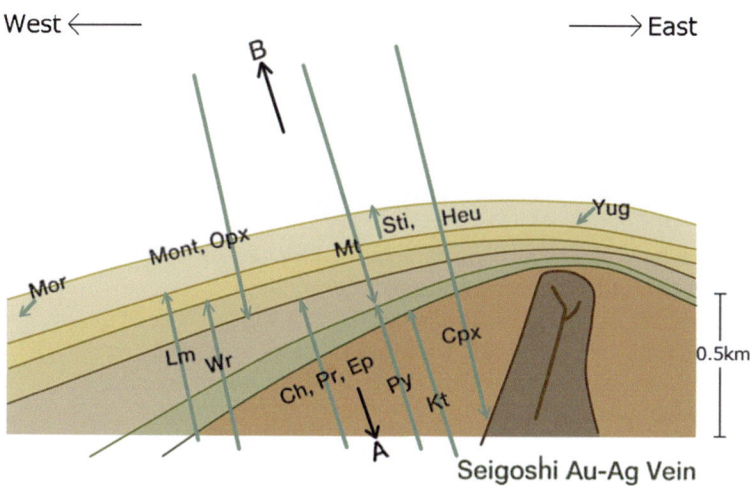

Fig. 1.19 Zonal sequence of the propylitic alteration in E–W section of the Seigoshi-Toi mine area, Shizuoka, Japan. *Yug* yugawaralite, *Heu* heulandite, *Stil* stilbite, *Opx* orthopyroxene, *Mont* montmorillonite, *Mor* mordenite, *Lm* laumontite, *Wr* wairakite, *Chl* chlorite, *pr* prehnite, *ep* epidote, *Py* pyrite, *Kf* K-feldspar, *Cpx* clinopyroxene (Shikazono 1985)

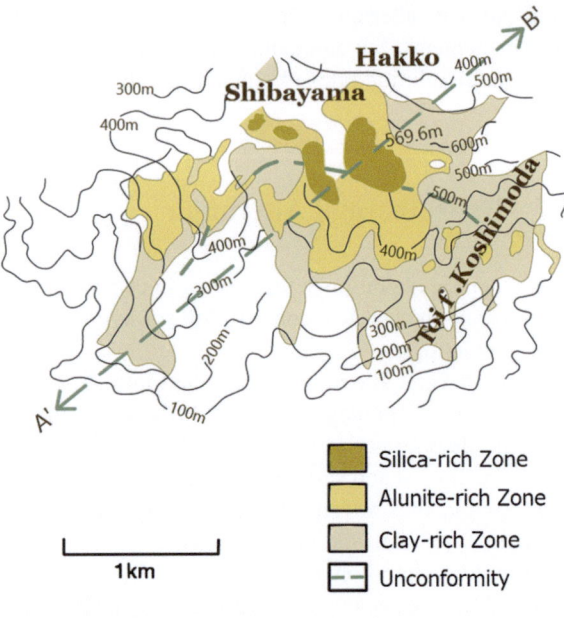

Fig. 1.20 Geology and alteration zoning in the Ugusu silica mine, Shizuoka, Japan. *Toi F* Toi Formation, *Koshimoda* Koshimoda andesite, *Hakko* Hakko orebody, *Shibayama* Shibayama ore body. Numbers indicate meters above sea level. Alteration zoning in the section of A′–B′ is shown in Fig. 1.21 (Shikazono 1985)

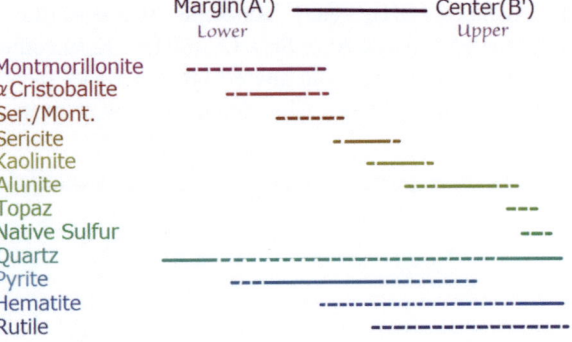

Fig. 1.21 Zonal sequence of the advanced argillic alteration from the central to marginal one in section of A′–B′ in Fig. 1.19 and from upper horizon to lower horizon (Shikazono 1985)

Usually the bulk chemical compositions of propylitically altered rocks are not distinctly different from those of original rocks.

Advanced argillic alteration is characterized by the presence of quartz, kaolinite ($Al_2Si_2O_5(OH)_4$), pyrophyllite ($Al_2Si_4O_{10}(OH)_2$), diaspore (α-AlOOH), and sericite ($KAl_3Si_3O_{10}(OH)_2$) which are formed by the interaction of rocks with strong acid hydrothermal solution.

The lateral and concentric alteration zonings are distinct for advanced argillic alteration (Figs. 1.20 and 1.21). Generally the alteration is caused by sulfuric acid solution. Many elements are leached out from the rocks under strong acid condition. SiO_2 content of central zone is high due to the low solubility of SiO_2 compared with other rock-forming minerals (e.g., feldspars) and addition of SiO_2 to the rocks from hydrothermal solution. Aluminum(Al) does not significantly remove from the rocks.

However, it removes and precipitates as alunite ($KAl_3(OH)_6(SO_4)_2$) at the margin of the central silica zone.

The strong acid hydrothermal solution is generated by the following reactions.

$$4SO_2(\text{volcanic gas}) + 4H_2O(\text{volcanic gas or ground water})$$
$$\rightarrow H_2S + 3H_2So_4 \tag{1.43}$$

$$H_2S(\text{volcanic gas}) + 2O_2 \rightarrow H_2SO_4 \tag{1.44}$$

1.8 Interpretation of Chemical Composition of Ground Water in Terms of Inverse Mass Balance Model

Meteoric water (rainwater) penetrates downward from the surface. It passes through unsaturated zone and becomes to ground water in saturated zone. Ground water reacts with surrounding rocks in saturated zone and reaches to chemical equilibrium condition after a long period. Chemical composition of ground water depends on the physicochemical and mineralogical properties of geologic media (rock type, cracks in rocks etc.). Ground water passes fractures in the rocks and emits from the surface, resulting to the source of surface water such as river water (Appendix, Plates 8 and 9). The reaction (e.g. dissolution of rocks) between surface water (rainwater, ground water) causes the change in surface topography (e.g. karst landscape (Appendix, Plate 10)).

Chemical compositions of ground water are plotted on H_4SiO_4 concentration (in logarithmic unit) − pH + pNa (pNa = $-\log m_{Na^+}$) diagram (Fig. 1.22).

As shown in Fig. 1.22, it can be generally said that the chemical compositions of shallow ground water vary widely. This wide variation suggests that the chemical equilibrium between minerals in rocks and shallow ground water is hard to be attained probably because of slow reaction rates of minerals at low temperatures, high flow rate of shallow ground water and short residence time. The chemical compositions of surface water are plotted in kaolinite region (Fig. 1.22). Therefore, if large amounts of surface water react with rocks, kaolinite, halloysite and allophane form.

The chemical compositions of deep ground water is not widely variable, compared with shallow ground water (Fig. 1.22). Flow rate is slow and residence time is long for deep ground water, resulting to the smaller deviation of chemical equilibrium between deep ground water and minerals contacting with deep ground water than that for shallow ground water.

Figure 1.23 shows the chemical compositions of ground water in Tono uranium mine area, Gifu, Central Japan which are plotted on the concentration diagram. The chemical compositions lie close to Ca-montmorillonite/kaolinite boundary (Yamakawa 1991). Residence time of the Tono ground water is estimated as ten thousands to several tens of thousands years from ^{14}C data. It is inferred that the chemical equilibrium was attained during such long time. On the other hand, the chemical compositions of shallow ground water in the Tono area is variable (Fig. 1.23).

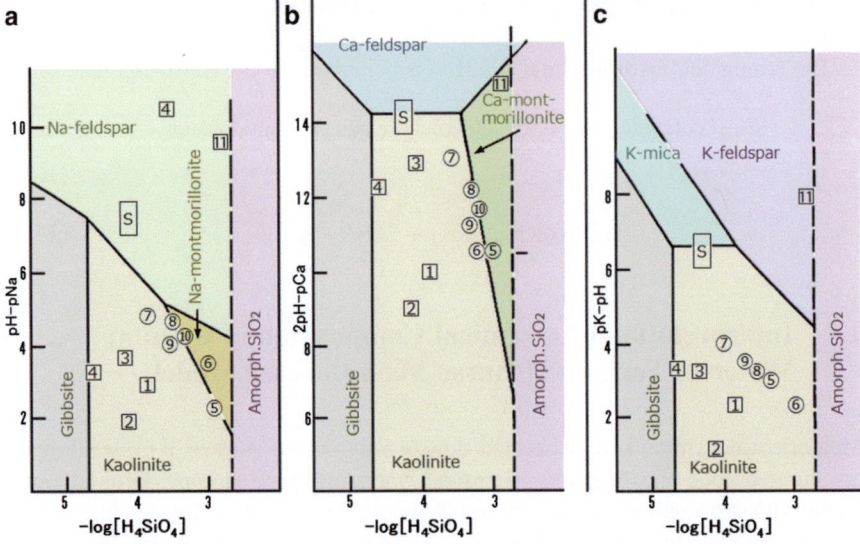

Fig. 1.22 Predominance diagrams illustrating the stability relations of some mineral phases (Stumm and Morgan 1970). In a typical weathering reaction, feldspars are converted (accompanied by increase of $[Na^+]/[H^+]$, $[Ca^{2+}]/[H^+]^2$ or $[K^+]/[H^+]$) into kaolinite. Progressive accumulation of H_4SiO_4 can lead to conversion of kaolinite into montmorillonite. For comparison, *squares* and *circles* represent analytical data of surface and ground waters, respectively; S represents seawater

Fig. 1.23 Chemical compositions of ground water and surface water in Tono area, Gifu, Japan (Yamakawa 1991). *Open triangle*: surface water, *filled square*: deep ground water, *cross, diamond, plus*: surface ground water

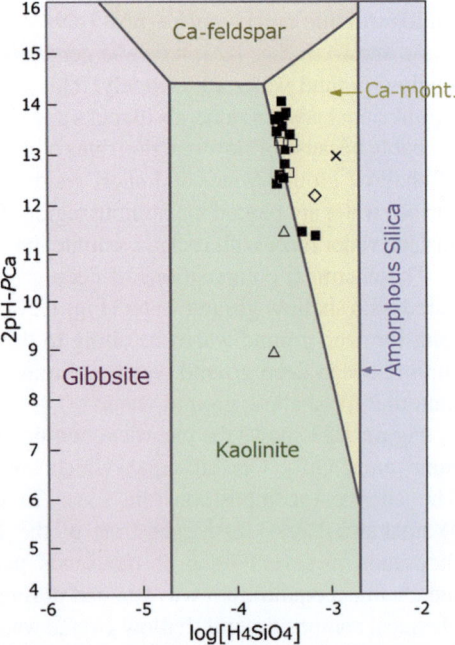

1.8 Interpretation of Chemical Composition of Ground Water in Terms of Inverse... 27

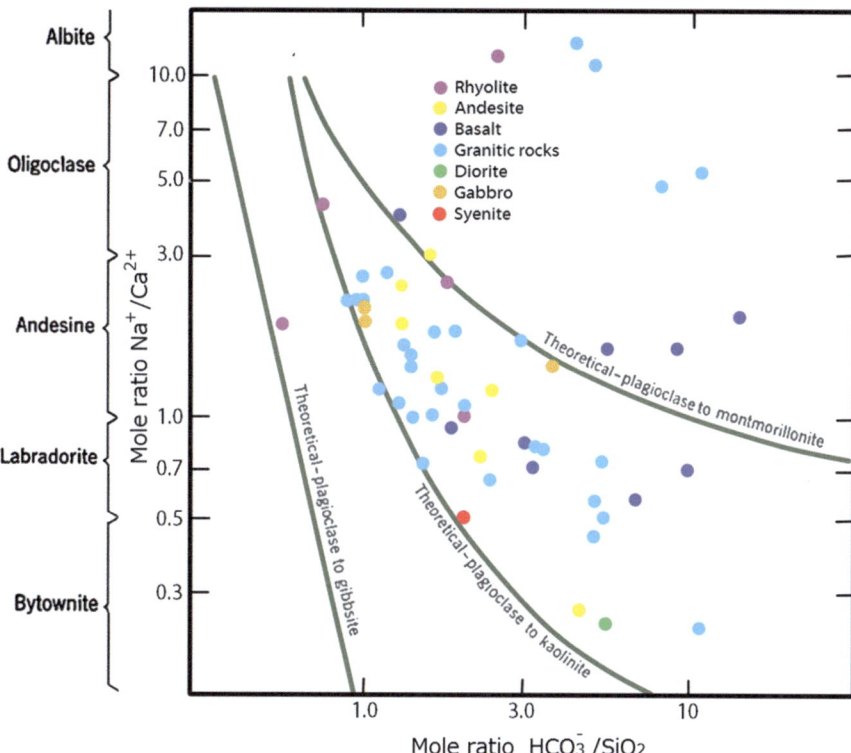

Fig. 1.24 Mole ratios of Na^+/Ca^{2+} plotted versus mole ratio of HCO_3^-/SiO_2 for waters from various types of igneous rocks. Most of the water compositions fall between the theoretical curve for waters from kaolinitization of plagioclase and the curve for montmorillonitization of plagioclase (Garrels 1967)

This wide variation may be caused by the short residence time and different extent of water–rock interaction.

The change in chemical composition of ground water interacting with feldspar which is the most common rock-forming mineral is considered below which is based on mass balance calculation in a similar manner of Garrels (1967).

Dissolution reaction of Na·Ca-feldspar (Na:Ca = 1:1 in atomic ratio) is written as

$$4Na_{0.5}Ca_{0.5}Al_{1.5}Si_{2.5}O_8(\text{NaCa-feldspar}) + 6CO_2 + 9H_2O$$
$$\rightarrow 3Al_2Si_2O_5(OH)_4(\text{kaolinite}) + 4SiO_2(\text{aqueous}) + 2Na^+ + 2Ca^{2+} + 6HCO_3^-$$
(1.45)

The ratios of SiO_2 (aqueous) to HCO_3^- and of Na^+ to Ca^{2+} dissolved from Na·Ca-feldspar by (1.45) are constant. However, these ratios differ for different Na·Ca-feldspars (Fig. 1.24). Similarly, the compositional change for the feldspar dissolution accompanied by smectite precipitation (Fig. 1.24) can be calculated. The calculated results for these dissolutions of feldspars accompanied by kaolinitization and montmorillonitization are shown in Fig. 1.24.

With the proceeding of these reactions, ground water becomes to be in equilibrium with kaolinite and Ca-montmorillonite. The reaction between kaolinite and montmorillonite is written as

$$7Al_2Si_2O_5(OH)_4(\text{kaolinite}) + Ca^{2+} + 2\,HCO_3^- + 8SiO_2(\text{aqueous})$$
$$= 6Ca_{0.17}Al_{2.34}Si_{3.66}O_{10}(OH)_2(\text{Ca-smectite}) + 2CO_2 + 9H_2O \quad (1.46)$$

Chemical equilibrium for (1.46) is given by

$$K_{1\text{-}36} = P_{CO_2}^2 / \left(a_{Ca^{2+}} a_{HCO_3^-}^{-2} a_{SiO_2}^8 \right) \quad (1.47)$$

CO_2 dissolves to aqueous solution by

$$CO_2 + H_2O = H^+ + HCO_3^- \quad (1.48)$$

Chemical equilibrium for this reaction is given by

$$K_{1\text{-}38} = (a_{H^+} a_{HCO_3^-})/P_{CO_2} \quad (1.49)$$

Using (1.47) and (1.49), it is derived that ($\log a_{Ca^{2+}} + 8\log a_{SiO_2} + 2pH$) is constant at constant temperature and pressure.

Compositional data on ground water are plotted on HCO_3^- concentration–$\log m_{Ca^{2+}} + 8\log m_{SiO_2} + 2pH$ diagram (Fig. 1.25). These data indicate that feldspar changes to kaolinite under the condition of low HCO_3^- concentration (less than 100 ppm) and ground water is nearly in equilibrium with kaolinite and smectite at high HCO_3^- concentration (more than 100 ppm). $\log m_{SiO_2}$ and ($\log m_{Ca^{2+}} + 2pH$) increase with the progress of feldspar dissolution accompanied by kaolinite precipitation. There are various trends of the composition of ground water during feldspar dissolution on $\log m_{SiO_2} - (\log m_{Ca^{2+}} + 2pH)$ diagram depending on P_{CO_2} (Fig. 1.26).

Ground water chemistry is controlled by several processes such as weathering, biological activity, mixing with other waters (e.g., soilwater, fossil seawater), anthropogenic influence and aerosol fall. The concentration of different element in ground water depends on different process. For instance, SiO_2 concentration totally depends on chemical weathering (dissolution of primary silicates and precipitation of secondary phases).

Approximately half of the amounts of alkali and alkali earth elements (K, Na, Ca, Mg) comes from dissolution of silicates, and the others from salt (evaporate) and carbonates. The concentrations of these elements in ground water can be interpreted based on dissolution and precipitation of silicates.

1.8 Interpretation of Chemical Composition of Ground Water in Terms of Inverse...

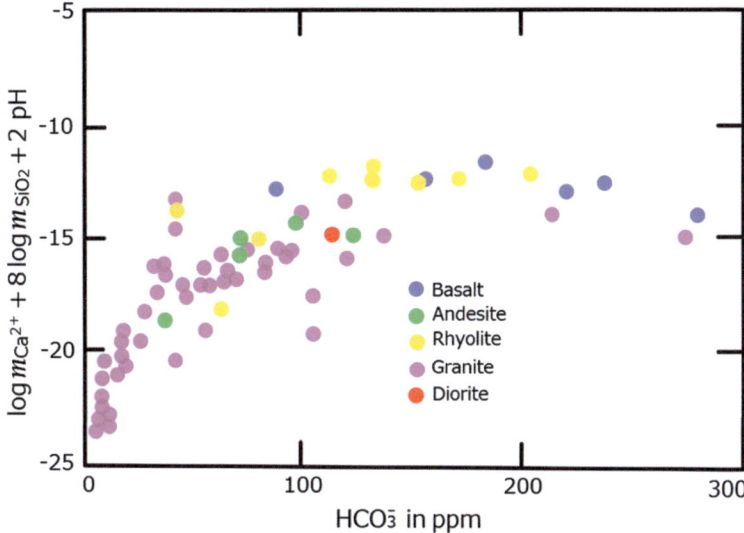

Fig. 1.25 Plot of ($\log m_{Ca^{2+}} + 8\log m_{SiO_2} + 2pH$) versus ppm HCO_3^- for waters from various igneous rocks. The tendency for constancy of the ordinate function at HCO_3^- concentration greater than about 100 ppm suggests approach to equilibrium between kaolinite and montmorillonite (Garrels 1967)

Mass balance simulation, which was used first by Garrels (1967), is useful to interpretating the ground water chemistry with regard to Si, alkali and alkali earth element concentrations.

Na·Ca-feldspar is the most common silicate mineral in earth's surface environment. It dissolves by the reaction

$$4Na_{0.5}Ca_{0.5}Al_{1.5}Si_{2.5}O_8(\text{NaCa-feldspar}) + 6H^+ + 11H_2O$$
$$\rightarrow 3Al_2Si_2O_5(OH)_4(\text{kaolinite}) + 4H_4SiO_4 + 2Na^+ + 2Ca^{2+} \qquad (1.50)$$

As shown in Fig. 1.26 the stable mineral in Na_2O–Al_2O_3–SiO_2–H_2O–CO_2 system in contacting with rainwater is kaolinite. Therefore, it is assumed that kaolinite precipitates according to reaction (1.50). It is also assumed that Na and Ca congruently dissolve from feldspar by reaction (1.50). This is experimentally indicated by several workers (e.g., Nesbitt et al. 1991). With the proceeding of reaction (1.50) from left hand side to right hand side, H^+ concentration decreases. However, H^+ is produced by the reactions

$$CO_2 + H_2O \rightarrow H_2CO_3 \qquad (1.51)$$

$$H_2CO_3 \rightarrow H^+ + HCO_3^- \qquad (1.52)$$

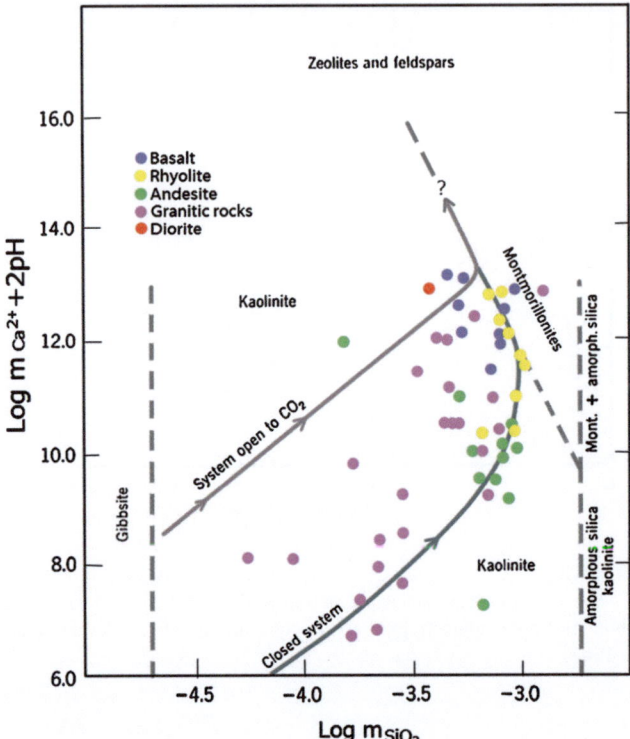

Fig. 1.26 Schematic diagram showing approximate solution compositions in equilibrium with one of the phases in the system $CaO–Al_2O_3–SiO_2–H_2O$, described in terms of Ca^{2+}, pH and SiO_2 (Garrels 1967). The *solid-line arrows* are calculated compositional changes with time for the alteration of a plagioclase ($Ab_{0.66}An_{0.34}$, Ab: albite component, An: anorthite component) by waters containing CO_2. The *straight arrow* path is based on the assumption that alteration took place in water maintained in equilibrium with the atmosphere ($P_{CO_2} = 10^{-3.5}$); the *curved arrow* path is calculated for a closed system with an initial dissolved CO_2 of 0.001 mol/L (an initial P_{CO_2} of $10^{-1.5}$ atm). Note that the composition of ground waters from igneous rocks are described better by the closed system path. *Dashed lines* are estimated solution compositions for phase boundaries

Assuming that the equilibrium is attained for (1.51) and (1.52), we obtain

$$K_{1\text{-}41} = m_{H_2CO_3}/P_{CO_2} \tag{1.53}$$

$$K_{1\text{-}42} = (m_{H^+} m_{HCO_3^-})/m_{H_2CO_3} \tag{1.54}$$

where m is molality, P_{CO_2} is partial pressure of CO_2 and $\gamma_{H_2CO_3^-}, \gamma_{H^+}, \gamma_{HCO_3^-}$, and γ_{CO_2} (γ: activity coefficient) is assumed to be unity.

From (1.53) and (1.54), we obtain

$$K_{1\text{-}42} = (m_{H^+} m_{HCO_3^-})/K_{1\text{-}41} P_{CO_2} \tag{1.55}$$

1.8 Interpretation of Chemical Composition of Ground Water in Terms of Inverse...

Electroneutrality relation in acid region is approximately given by

$$2m_{Ca^{2+}} + m_{Na^+} + m_{H^+} = m_{HCO_3^-} \tag{1.56}$$

From (1.56), and $m_{Na^+} = m_{Ca^{2+}}$, we obtain

$$3m_{Ca^{2+}} + m_{H^+} = m_{HCO_3^-} \tag{1.57}$$

It is assumed above that dominant cation and anion species are Ca^{2+}, Na^+, H^+ and HCO_3^-.

From (1.55) and (1.57), we obtain

$$m_{H^+}^2 + 3m_{Ca^{2+}}m_{H^+} - K_{1\text{-}41}K_{1\text{-}42}P_{CO_2} = 0 \tag{1.58}$$

This relation is solved as

$$pH = -\log\left\{-3/2 m_{Ca^{2+}} + 9m_{Ca^{2+}} + 4K_{1\text{-}41}K_{1\text{-}42}P_{CO_2}\right)^{1/2}\right\} \tag{1.59}$$

This equation means that pH is determined, if temperature, P_{CO_2} and $m_{Ca^{2+}}$ are given.

If initial solution composition is same to that of average rainwater in Japan (Nishimura 1991; Si 0.83 ppm, Na 1.1 ppm, Ca 0.97 ppm), the initial composition can be plotted on $(\log m_{Ca^{2+}} + 2pH) - \log m_{H_4SiO_4}$ diagram.

It is considered here that Δ mol feldspar reacts with 1 kg rainwater with the above initial composition. After this reaction, the chemical compositions of solution at step 1 are given by

$$\left.\begin{array}{l}(m_{Ca^{2+}})_1 = (m_{Ca^{2+}})_i + (1/2)\Delta \\ (m_{Na^+})_1 = (m_{Na^+})_i + (1/2)\Delta \\ (m_{H_4SiO_4})_1 = (m_{H_4SiO_4})_i + \Delta\end{array}\right\} \tag{1.60}$$

where i is initial.

From (1.60) and giving Δ, m_1 is obtained. Putting this m_1 to (1.59), we obtain pH at constant P_{CO_2}. Stepwise calculations yield the compositional trend during the feldspar dissolution on $(\log m_{Ca^{2+}} + 2pH) - \log m_{H_4SiO_4}$ diagram (Fig. 1.26) and also on $(\log m_{Na^+} + pH) - \log m_{H_4SiO_4}$ diagram.

Δ corresponds to ς (degree of reaction) which is used by Prigogine (1967) in the field of irreversible thermodynamics.

Helgeson (1971, 1972, 1974) applied irreversible thermodynamics using ς to the compositional variations in aqueous solution reacting with K-feldspar. Above arguments consider only feldspar dissolution. However, other silicate minerals (olivine, pyroxene etc.) react with ground water. For example, Mg-pyroxene and Mg-olivine dissolve by

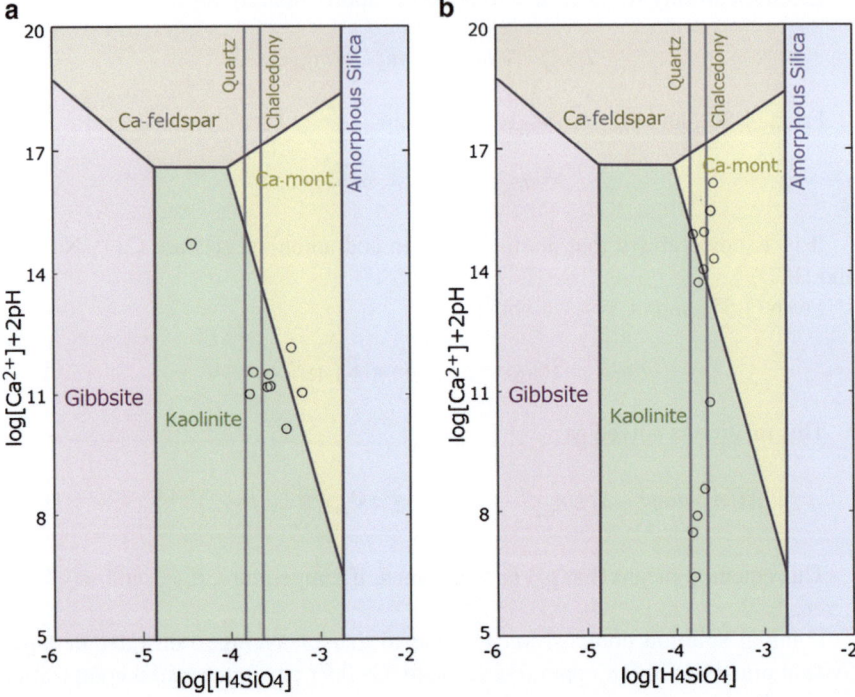

Fig. 1.27 Chemical composition of ground water in granitic rock area plotted on $\log m_{Ca^{2+}} + 2\text{pH} - \log m_{H_4SiO_4}$ diagram. (**a**) Shallow ground water (Tsukuba, Japan), (**b**) Deep ground water (Kamaishi mine, Japan)

$$\left. \begin{array}{l} \text{MgSiO}_3(\text{Mg-pyroxene}) + 2\text{H}^+ + \text{H}_2\text{O} \rightarrow \text{Mg}^{2+} + \text{H}_4\text{SiO}_4 \\ \text{Mg}_2\text{SiO}_4(\text{Mg-olivine}) + 4\text{H}^+ \rightarrow 2\text{Mg}^{2+} + \text{H}_4\text{SiO}_4 \end{array} \right\} \quad (1.61)$$

H_4SiO_4 concentration increases with the proceeding of these reactions. Therefore, it is obvious that these reactions affect the compositional trend on $(\log m_{Ca^{2+}} + 2\text{pH}) - \log m_{H_4SiO_4}$ diagram.

Several investigations on the change in chemical composition of ground water interacting with rocks composed of several minerals have been applied to the interpretation of ground water chemistry. For example, Garrels and Mackenzie (1967) interpreted ground water chemistry in granitic rock area in Shela Nevada, U.S.A. based on mass balance method mentioned above. Computer codes including not only acid–base reactions which are considered in mass balance method, but also oxidation-reduction reaction, isotope mass balance and mixing of fluids have been developed and can be used to interpret geochemistry of ground water (BALANCE: Parkhurst et al. 1990; NETPATH: Plummer et al. 1991).

Analytical data on ground water in granitic rock areas (Japan, U.S.A., Canada, Sweden etc.) are plotted on $\log m_{Ca^{2+}} + 2\text{pH} - \log m_{H_4SiO_4}$ diagram (Fig. 1.27).

The chemical features of the ground water in granitic rock areas are

(1) All analytical data plot in kaolinite and Ca-montmorillonite region and no data in gibbsite region.
(2) Data plotted in kaolinite region indicate that $\log m_{H_4SiO_4}$ increases with increasing ($\log m_{Ca^{2+}} + 2pH$). This trend seems to be consistent with the prediction by mass balance method mentioned above, but the slope seems higher than the theoretical one, indicating that dissolution of $Na_{0.5}Ca_{0.5}$-feldspar cannot explain the variation in analytical data. This discrepancy may be due to the precipitation of chalcedony and quartz. Some data are plotted close to the saturation lines of chalcedony and quartz.
(3) H_4SiO_4 concentration of ground water from Strippa mine (Sweden), Canadian Shield (Canada) and Kamaishi Cu–Fe mine (Japan) are plotted along these lines.
(4) Some data are plotted along kaolinite/Ca-montmorillonite line (ground water in USA) and in Ca-montmorillonite region (Abukuma river, Fuji river; Japan).
(5) Some ground water data from Abukuma river, Fuji river and Strippa regions are plotted along kaolinite-Ca-montmorillonite boundary line and close to Ca-feldspar stable region. These ground waters are considered to be highly evolved ground water.

1.9 Oxidation-Reduction Condition

1.9.1 Oxygen Fugacity (f_{O_2})-pH Diagram

Oxidation-reduction reactions are important for the elements with various charges (e.g., S, Fe, Mn) in aqueous solution. Stability relations for these elements in aqueous solution are derived using equilibrium constants in Table 1.4.

1.9.2 H–S–O System

Stability condition of aqueous sulfur species in H–S–O system will be considered below (Barnes and Kullerud 1961). It is assumed that dominant aqueous sulfur species are H_2S, HS^-, S^{2-} and SO_4^{2-} and the concentrations of other aqueous sulfur species (e.g., $S_2O_3^{2-}$, SO_3^{2-}, HSO_3^-, $HS_2O_3^-$, S_2^{2-}, S_3^{2-}, S_4^{2-}, H_2S_x, HS_x^-, S_x^{2-}) are neglected. The reactions among the aqueous sulfur species are given by

$$H_2S = HS^- + H^+ \tag{1.62}$$

Table 1.4 Equilibrium constants (log K values) for some hydrolysis and redox reactions (data from Helgeson et al. 1978; Robie et al. 1978; and Fisher and Barnes 1973), except for reaction (2)

	Reaction	25	50	100	150	200	250	300	350 °C	a	b
1.	$HSO_4^- = H^+ + SO_4^{2-}$	−1.99	−2.31	−2.99	−3.72	−4.48	−5.27	−6.08	−6.90	−13.59	4,324.08
2.	$H_2S_{aq} = HS^- + H^+$	−6.98	−7.72	−6.61	−6.81	−7.17	−7.60	−8.05	–	−12.18	2,377.5
3.	$HS^- + 2O_2 = SO_4^{2-} + H^+$	132.55	120.43	100.91	85.86	73.82	63.90	55.38	47.28	−31.75	4,995.5
4.	$3Fe_2O_3 = 2Fe_3O_4 + 1/2O_2$	−36.15	−32.82	−27.51	−23.45	−20.24	−17.65	−15.51	−13.70	6.9	−12,826.3
5.	$3Fe + 2O_2 = Fe_3O_4$	177.40	162.37	138.21	119.78	105.27	93.56	83.91	75.83	−17.13	57,893.3
6.	$3FeS + 3H_2O + 1/2O_2 = Fe_3O_4 + 3HS^- + 3H^+$	−6.14	−6.41	−7.34	−8.64	−10.28	−12.08	−14.71	−19.17	−35.25	11,905.6
7.	$FeS + H_2S_{(g)} + 1/2O_2 = FeS_2 + H_2O$	46.09	41.54	34.29	28.77	24.40	20.88	17.98	15.55	−12.4	17,410.7
8.	$Fe^{2+} + 2H_2S_{(g)} + 1/2O_2 = FeS_2 + 2H^+ + H_2O$	30.24	27.45	23.07	19.80	17.32	15.48	14.37	14.88	0.263	8,038.5
9.	$Fe^{2+} + 3H_2O = Fe(OH)_3 + 3H^+$	−29.45	−27.58	−24.64	−22.42	−20.70	−19.33	−18.21	–	−6.42	−6,753.56
10.	$H_2O_{(l)} = H_2 + 1/2O_2$	−41.55	−37.68	−31.53	−26.85	−23.18	−20.23	−17.81	−15.78	7.6	−14,564.13
11.	$H_2O_{(l)} = H^+ + OH^-$	−13.99	−13.26	−12.23	−11.59	−11.21	−11.08	−11.28	−12.35	–	–
12.	$1/2H_2 = H^+ + e^-$	0.0	0.0	0.0	0.0	0.0	0.0	0.0	0.0	–	–
13.	$2H_2S + O_{2(g)} = S_{2(g)} + 2H_2O$	59.44	54.4	46.27	39.99	35.05	30.06	27.21	–	−9.99	21,287.7
14.	Alunite + quartz + H_2O = K-mica + kaolinite + 2 K^+ + 6H^+ + 8HSO_4^-	–	–	–	–	−48.0	−41.7	−36.5	–	17.90	−31,168
15.	$N_2 + 3H_2 = 2NH_3$	5.75	4.48	2.44	0.86	−0.42	−1.46	−2.34	−3.09	−11.41	5,201.7
16.	$C + O_2 = CO_2$	69.09	63.79	55.26	48.74	43.60	39.45	36.02	33.13	0.168	20,543.6
17.	$C + 2H_2 = CH_4$	8.88	7.86	6.20	4.92	3.89	3.04	2.32	1.71	−5.10	4,253.4
18.	$NH_4^+ = NH_3 + H^+$	−9.27	−8.53	−7.42	−6.53	−5.79	−5.14	−4.57	−4.12	−1.19	3,303.4

a and b are linear regression equations for these data: log K = a + b (l/T K) (range: 200–300 °C) (Barton 1984)

1.9 Oxidation-Reduction Condition

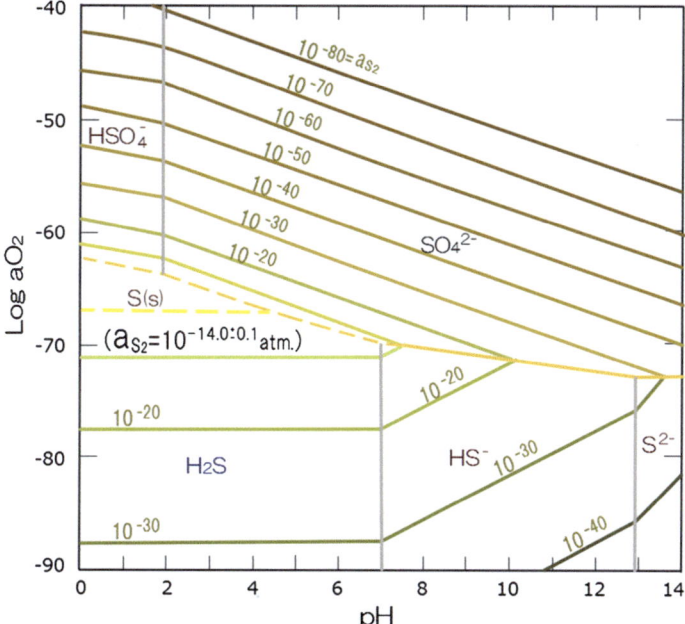

Fig. 1.28 Stabilities of aqueous sulfur-containing species at $\sum S$ (total dissolved S concentration) = 0.1 mol/kg · H_2O, 25 °C. (**a**) Light, solid contours show a_{S_2}. Changes in the fields of predominance (heavy, *solid lines*) on decreasing $\sum S$ to 0.001 are indicated by the heavy, *dashed lines* (Barnes and Kullerud 1961)

$$HS^- = H^+ + S^{2-} \quad (1.63)$$

$$H_2S + 2O_2 = SO_4^{2-} + 2H^+ \quad (1.64)$$

$$HSO_4^- = H^+ + SO_4^{2-} \quad (1.65)$$

Mass balance equation for dissolved S is written as

$$\sum S = m_{H_2S} + m_{HS^-} + m_{S^{2-}} + m_{SO_4^{2-}} + m_{HSO_4^-} \quad (1.66)$$

where $\sum S$ is total dissolved sulfur concentration.

Using this equation and mass action law for Eqs. (1.62), (1.63), (1.64), and (1.65), stability fields of aqueous sulfur species can be shown on $\log f_{O_2}$-pH diagram at constant temperature, pressure and ionic strength. The boundary lines for dissolved sulfur species on Fig. 1.28 correspond to the concentration of each aqueous sulfur species which is equal to that of the other one. For example, at the boundary line between H_2S and HS^-, m_{H_2S} is equal to m_{HS^-}. pH at this boundary can be obtained from equilibrium constant for (1.52) which is given by

$$K_{1\text{-}52} = (a_{HS^-}a_{H^+})/a_{H_2S} = (\gamma_{HS^-}m_{HS^-}a_{H^+})/(i_{H_2S}m_{H_2S}) = (\gamma_{HS^-}a_{H^+})/\gamma_{H_2S} \quad (1.67)$$

Therefore, pH is represented by

$$pH = -\log K_{1\text{-}40} + \log(\gamma_{HS^-}/\gamma_{H_2S}) \quad (1.68)$$

1.9.3 Fe–S–O–H System

The thermochemical stability of solid phases, aqueous species and gaseous species in Fe–S–O–H system will be considered below. Solid phases of this system include Fe, FeO, Fe_3O_4, Fe_2O_3, FeS, and FeS_2. Dominant aqueous species considered are H_2S, HS^-, S^{2-}, HSO_4^-, SO_4^{2-}, Fe^{2+}, Fe^{3+}, $Fe(OH)^+$, $Fe(OH)_2$ and $Fe(OH)_3$. Gaseous species are S_2, O_2, and H_2O. The chemical equilibria among these phases and dissolved species are assumed. For instance, the following reactions are taken into account.

$$FeS + (1/2)S_2 = FeS_2 \quad (1.69)$$

$$2FeS_2 + (3/2)O_2 = Fe_2O_3 + 2S_2 \quad (1.70)$$

$$2Fe_3O_4 + (1/2)O_2 = 3Fe_2O_3 \quad (1.71)$$

From the equilibrium constants of (1.69), (1.70) and (1.71), we obtain the stability fields of iron minerals on $\log f_{O_2}$-pH diagram at constant temperature, pressure and $\sum S$ (Fig. 1.29) and on Eh (oxidation-reduction potential)-pH diagram.

1.9.4 Estimate of Oxygen Fugacity (f_{O_2}) for Hydrothermal Ore Deposits

Fe-minerals and sulfides are common in hydrothermal ore deposits. If we know these mineral species, chemical compositions of minerals and temperature, f_{O_2}-pH and Eh-pH conditions can be restricted. As an example of the estimates of f_{O_2} and pH, vein-type deposits in Japan are taken into account. The vein-type deposits considered here include precious metal (Au, Ag) and base metal (Cu, Pb, Zn, Fe, Mn) deposits and are classed as epithermal-type (Shikazono 2003). The oxidation-reduction condition can be estimated based on mineral assemblages and chemical compositions of minerals. Sphalerite is the most common sulfide mineral in these deposits. It contains iron and is zinc–iron solid solution ($Zn_{1-x}Fe_xS$). The relationship among FeS content of sphalerite, f_{O_2}, pH and temperature is obtained from the reaction.

1.9 Oxidation-Reduction Condition

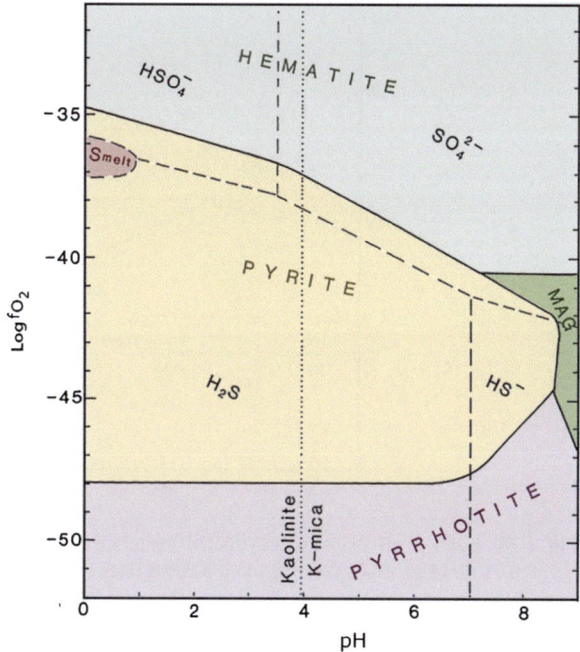

Fig. 1.29 Log f_{O_2}-pH diagram for Fe–S–O–H system at 200 °C. The K-mica-kaolinite curve is taken from Hemley et al. (1969) Log a_{K^+} is -1.25 (Barton 1984)

$$(FeS)_{sphalerite} + H_2S + 1/2O_2 = FeS_2(pyrite) + H_2O \quad (1.72)$$

$$H_2S + 2O_2 = SO_4^{2-} + 2H^+ \quad (1.73)$$

Equilibrium constants of (1.72) and (1.73) are given by the following Eqs. (1.74) and (1.75).

$$K_{1\text{-}62} = 1/\left(a_{FeS} a_{H_2S} f_{O_2}^{1/2}\right) \quad (1.74)$$

where a_{FeS_2} and a_{H_2S} are activity of FeS component in sphalerite and of H_2S, respectively, a_{FeS_2} and a_{H_2O} are assumed to be unity.

$$K_{1\text{-}63} = \left(a_{SO_4^{2-}} a_{H^+}^2\right)/\left(a_{H_2S} f_{O_2}^2\right) \quad (1.75)$$

$\sum S$ (total dissolved sulfur concentration) is approximated as

$$\sum S = m_{H_2S} + m_{HS^-} + m_{SO_4^{2-}} \quad (1.76)$$

From three Eqs. (1.74, 1.75, and 1.76), iso-FeS contours are drawn on log f_{O_2}-pH diagram at constant temperature, $\sum S$, and ionic strength (Fig. 1.30).

Many analytical data on sphalerite from epithermal precious and base metal vein-type deposits in Japan indicate that the FeS contents of sphalerite from

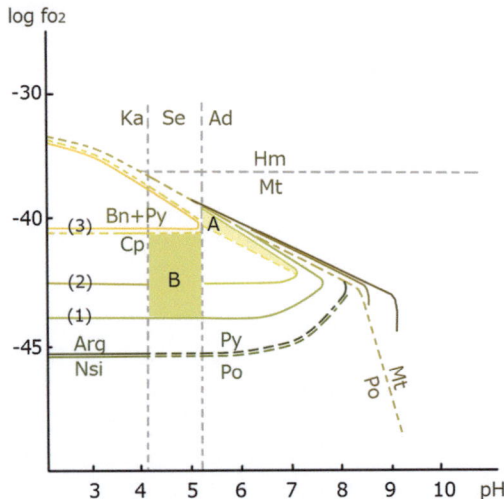

Fig. 1.30 Log f_{O_2}-pH diagram constructed for temperature = 250 °C, ionic strength = 1, and $\sum S = 0.01$ mol/kg· H_2O. Iso-FeS mole percent lines for sphalerite and the stability relations of some hydrothermal minerals are also given. The *solid lines* (*1*), (*2*) and (*3*) are 10, 1 and 0.1 mol% FeS in sphalerite, respectively; The *dotted* and *shaded areas* of *A* and *B* represent the possible f_{O_2}-pH regions for Japanese Au–Ag vein-type and Pb–Zn vein-type deposits, respectively. *Bn* bornite, *Py* pyrite, *Cp* chalcopyrite, *Arg* argentite, *Nsi* native silver, *Hm* hematite, *Mt* magnetite, *Po* pyrrhotite, *Ser* sericite, *Ad* adularia (Shikazono 1978a, b)

precious vein-type and base metal vein-type deposits are 1 mol% and 1–20 mol%, respectively (Shikazono 1974). Using the FeS content of sphalerite, mineral assemblage, estimated temperature and salinity and $\sum S$, f_{O_2}-pH conditions for the two types of deposits were restricted (Fig. 1.30). The difference in f_{O_2} and pH conditions for these deposits is considered to be caused by boiling of fluids, interaction of hydrothermal solution with host rocks, mixing of hydrothermal solution with ground water and temperature. More detailed discussions on the chemical environments of these two types of ore depositions are given in Shikazono (1978a, b, 2003).

1.9.5 Sulfides

Sulfides are usually abundant in metallic ore deposits. Ore zoning is observed frequently in the ore district. For example, ore district zoning away from granitic intrusive body is Sn·W-deposits, Cu-deposits, Zn·Pb-deposits and Au·Ag-deposits. Figure 1.31 shows ore body zoning in vein-type deposits from lower to upper horizons, that is, Cu-zone → Zn·Pb-zone → Au·Ag-zone. These zonings are considered to reflect solubilities of ore minerals (Barnes 1979). In an ideal case, the zonal sequence away from the intrusive igneous body is consistent with the order of

1.9 Oxidation-Reduction Condition

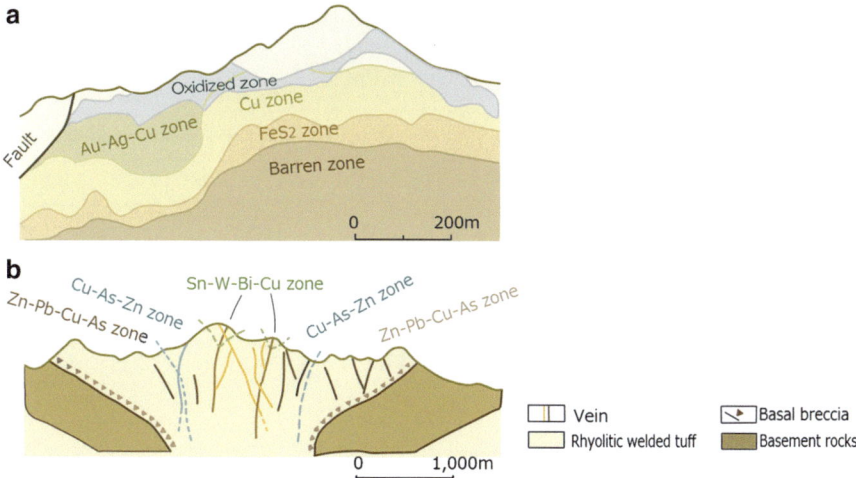

Fig. 1.31 Metal zoning in Osarizawa ore deposit (Saisei-hi, cross section, (**b**)) and Ashio deposit, cross section, (**a**) (Horikoshi 1977)

solubility of ore minerals. Low solubility minerals tend to precipitate near the intrusive body, while high solubility minerals far away from the intrusive body, although the zonal sequence of ore minerals depends on concentrations in ore fluids, temperature, and the dependence of solubility on temperature.

Solubilities of sulfides in pure water are very low, but they increase significantly in saline and H$_2$S-bearing aqueous solutions. This is due to the stable existence of complexes such as chloro- and thio-complexes. The solubilities of sulfides strongly depend on stability constants of these complexes. Sulfides precipitate when saturation index (S.I.) (S.I. is defined as S.I. = log(I.A.P/K$_{sp}$, where I.A.P is activity product and K$_{sp}$ is solubility product.) exceeds zero during the flow of ore fluids accompanied by the changes in physico-chemical variables (e.g., decrease of temperature). As an example, precipitation of PbS is considered below. Ionic activity product for PbS (I.A.P) is expressed as

$$\text{I.A.P} = a_{Pb^{2+}} a_{S^{2-}} \tag{1.77}$$

If S.I. changes with the changes in Pb^{2+} concentration, S^{2-} concentration and salinity, resulting to more than zero of the S.I., PbS precipitates. Solubility of PbS in Pb–S–Cl–H$_2$O system is derived below in order to consider the precipitation mechanism for PbS (Henley 1984).

$$\text{PbS} + 2\text{H}^+ = \text{Pb}^{2+} + \text{H}_2\text{S} \tag{1.78}$$

$$\text{Pb}^{2+} + \text{Cl}^- = \text{PbCl}^+ \tag{1.79}$$

$$\text{PbCl}^+ + \text{Cl}^- = \text{PbCl}_2 \tag{1.80}$$

$$PbCl_2 + Cl^- = PbCl_3^- \qquad (1.81)$$

$$PbCl_3^- + Cl^- = PbCl_4^{2-} \qquad (1.82)$$

If $PbCl_2$ is the most dominant among dissolved Pb species, solubility of PbS can be derived as follows.

$$PbS + 2H^+ + 2Cl^- = PbCl_2 + H_2S \qquad (1.83)$$

Equilibrium constant for this reaction is expressed as

$$K_{1\text{-}73} = (a_{PbCl_2} a_{H_2S})/(a_{H^{+2}} a_{Cl^-}^2) \qquad (1.84)$$

Thus, we obtain

$$m_{PbCl_2} = \left(K_{1\text{-}73} m_{Cl^-}^2 a_{H^+}^2 \gamma_{Cl^-}^2\right) / \left(m_{H_2S} \gamma_{H_2S} \gamma_{PbCl_2}\right) \qquad (1.85)$$

Equation (1.85) indicates that the concentration of $PbCl_2$ which is approximated as total dissolved Pb depends on temperature, total pressure, m_{Cl^-}, pH, m_{H_2S} and salinity.

Next, we consider the solubility of sulfide assemblage (chalcopyrite + bornite + pyrite). This is determined by the reaction

$$5CuFeS_2(\text{chalcopyrite}) + S_2 = Cu_5FeS_4(\text{bornite}) + 4FeS_2(\text{pyrite}) \qquad (1.86)$$

It is obvious that the chemical equilibrium for this reaction depends on temperature, f_{S_2} and total pressure. It is derived from the equilibrium relation from the following reaction that shows f_{S_2} relates to f_{O_2}, m_{H_2S}, temperature and total pressure.

$$S_2 + H_2O = H_2S + 1/2O_2 \qquad (1.87)$$

From the reactions (1.86) and (1.87) and chemical equilibrium involving Cu- and Fe-chloro-complexes, iso-concentration contours for Cu- and Fe-minerals are drawn on $\log f_{O_2}$-pH diagram (Fig. 1.32; Crerar and Barnes 1976). Above calculations were carried out, assuming that Cu- and Fe-chloro-complexes are dominant Cu- and Fe-species (Crerar and Barnes 1976). Bisulfide-, thio- and carbonate-complexes are also important in determining solubility of sulfides as well as chloro-complexes (Table 1.5, Brimhall and Crerar (1987)). Above argument clearly indicates that dominant base metal complexes in hydrothermal solution depend on concentration of ligand, pH, f_{O_2}, temperature and so on.

Solubility of sulfides largely depends on stability of base metal complexes. The stability of complexes is determined by (1) hardness and softness of acids–bases (HSAB), (2) ligand-field stabilization energy (LFSE), and (3) temperature, and total pressure.

1.9 Oxidation-Reduction Condition

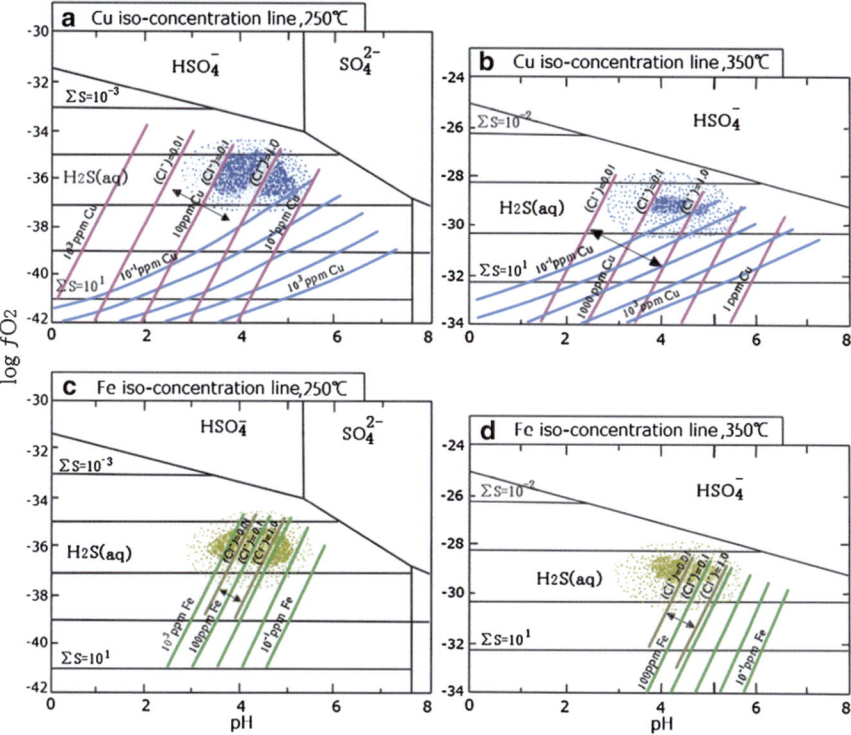

Fig. 1.32 Cu and Fe concentration in hydrothermal solution in equilibrium with chalcopyrite-bornite-pyrite assemblage (Crerar and Barnes 1976). *Solid line*: solubility due to chloro complexes, *dotted line*: solubility due to bisulfide complexes, *dotted region*: f_{O_2}-pH range of porphyry copper deposits

Table 1.5 Dissolved species of metal elements in hydrothermal solution (Crerar et al. 1985)

Metal	Low Cl$^-$ concentration		High Cl$^-$ concentration
	Low pH	High pH	
Fe	Fe^{2+}	FeOH$^+$, Fe(OH)$_2$	FeCl$_n^{2-n}$(n = 0–3)
Zn	Zn^{2+}	ZnOH$^+$, Zn(OH)$_2$	ZnCl$_n^{2-n}$(n = 0–3)
Pb	Pb^{2+}	PbOH$^+$, Pb(OH)$_2$	PbCl$_n^{2-n}$(n = 0–3)
Bi	Bi^{2+}	BiOH$^+$, Bl(OH)$_2$	BiCl$_n^{2-n}$(n = 0–2)
Au	AuHS or HAu(HS)$_2$	Au(HS)$_2^-$	AuCl
Ag	AgHS or HAg(HS)$_2$	Ag(HS)$_2^-$	AgCl
Mo	H$_2$MoO$_4$ HMoO$_4^-$	MoO$_4^{2-}$	not variable
Mo	MoHCO$_3^-$ or CO$_3^{2-}$ complex		
Sb	Sb(aq), HSbS$_2$, H$_2$SbS$_4$, Sb(OH)$_2^+$, Sb(OH)$_3$		not variable

Table 1.6 Classification of cations and anions by hardness and softness by Pearson (1968), Crerar et al. (1985)

Hard acids	Mn^{2+}, Ga^{3+}, In^{3+}, Co^{2+}, Fe^{3+}, As^{3+}, Sn^{4+}, MoO^{3+}, WO^{4+}, CO_2
Soft acids	Cu^+, Ag^+, Au^+, Ti^+, Hg^+, Cd^{2+}, Hg^{2+}, Te^{4+}, Tl^{3+}
Intermediate acids	Fe^{2+}, Co^{2+}, Ni^{2+}, Cu^{2+}, Zn^{2+}, Pb^{2+}, Sn^{2+}, Sb^{3+}, Bi^{3+}, SO_2
Hard bases	OH^-, Cl^-, CO_2, SO_2
Soft bases	H_2S, HS^-, S^{2-}
Relative hardness of metal ions and ligands	
$F^- > Cl^- > Br^- > I^-$	$Zn^{2+} > Pb^{2+}$
$Cu^+ > Ag^+ > Au^+$	$H^+ > Li^+ > Na^+ > K^+ > Rb^+ > Cs^+$
$Zn^{2+} > Cd^{2+} > Hg^{2+}$	$As^{3+} > Sb^{3+} = Bi^{3+}$

(1) Hardness and softness of acids and bases

HSAB (Hard and Soft, Acids and Bases) principle was proposed by Pearson (1963, 1968). This is summarized as (1) hard acids are in an affinity with hard bases, (2) soft acids are in large volume and have small charge and their electronegativity is large, (3) hard acids react with hard acids to form ionic compounds, while (4) covalent compounds form by the reaction of soft acids with soft bases. Acids and bases are classified according to this principle (Table 1.6). For instance, the order of hardness is $F^- > Cl^- > Br^- > I^-$, $Cu^+ > Ag^+ > Au^+$. HS^- tends to combine with soft cations such as Au^+, Hg^+ and Pt^+ to form HS^- complexes. Base metal cations such as Zn^{2+}, Pb^{2+}, Cu^+, and Fe^{2+} tend to combine with relatively hard anions such as Cl^- to form chloro-complexes.

This concept of HSAB can be applied to interpret the difference in the base metals concentrated to different vein-type deposits in Japan (Shikazono and Shimizu 1992).Major vein-type deposits in Japan include precious metal deposits and base metal deposits which is characterized by enrichments of Hg, Te, Se, As, Sb, Cd and Tl in addition to Au and Ag, and Cu, Pb, Zn, Fe, Mn, Ag, Cu, Bi, As, Sb, In, Ga, Sn and W, respectively. This means precious metal and base metal vein-type deposits are characterized by enrichment of soft metals and hard metals, respectively. This difference is consistent with fluid inclusion data showing that Cl^- of ore fluids responsible for base metal deposits is distinctly higher than that responsible for precious deposits (Shikazono and Shimizu 1992).

(2) Ligand field stabilization energy (LFSE)

d orbital of transition metals in octahedral ligand field splits into tzg orbital with low energy and e_g with high energy. Example of this ligand field splitting for d6 ion (Fe^{2+}, Sc^{2+} etc.) is shown in Fig. 1.33. Energy of splitting orbital is 4Dqoct lower than the normal orbital. This energy is called LFS (ligand field stabilization energy).

Figure 1.34 shows the relationship between LFSE and z_i/r_i (z: charge, r_i: ionic radii). From this figure metallic elements are divided into two types: elements occurring in common hydrothermal ore deposits and in rare hydrothermal ore deposits.

1.9 Oxidation-Reduction Condition

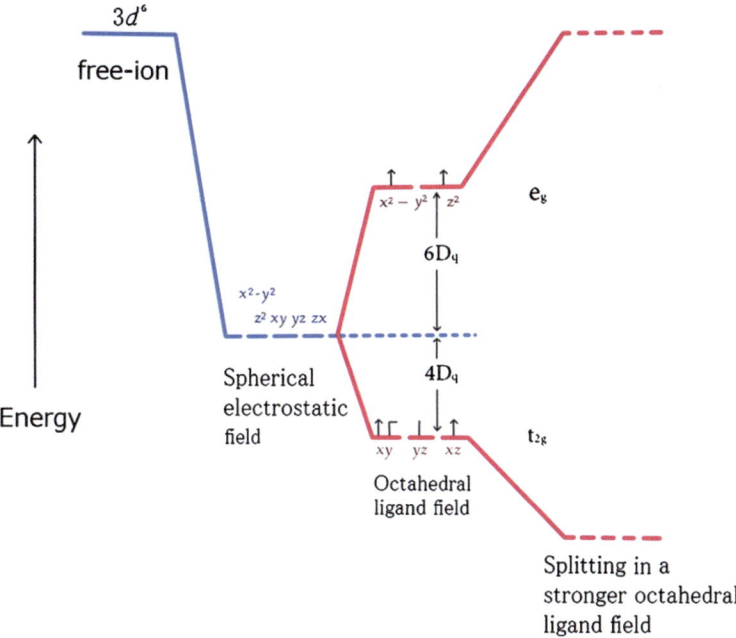

Fig. 1.33 Effect of an applied octahedral ligand-field on the total energies of 3d complex ions using high spin Fe^{2+} ($3d^6$) as an example (Crerar et al. 1985)

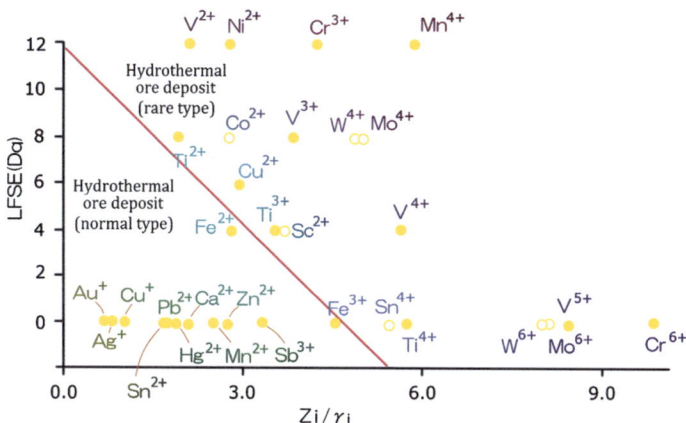

Fig. 1.34 Relation between LFSE (Ligand Field Stabilization Energy) and z_i (charge)/r_i (ionic radii) (Crerar et al. 1985)

(3) Temperature and total pressure

Temperature significantly influences on stability of complexes. The structure of aqueous solution changes according to the change in temperature. Chlorocomplexes change from octahedral coordination to tetrahedral coordination with increasing temperature by the following reaction (Crerar et al. 1985; Uchida et al. 1996).

$$MCl_n(OH_2)^{6-n}(\text{octahedral}) = MCl_x(OH_2)^{4-x}(\text{tetrahedral}) \tag{1.88}$$

where M is metallic element.

Susak and Crerar (1985) inferred that the base metals solubility at elevated temperatures is due to complexes with tetrahedral coordination, indicating that metals are transported as chloro-complexes at elevated temperatures and ore minerals precipitate when octahedral coordination changes to tetrahedral coordination with decreasing temperature. Temperature largely influences on dielectric constant of water as well as coordination. The dielectric constant of water at 25 °C and vapour saturated pressure condition is 78.47 and it decreases to 12.87 at 350 °C. This implies that ionic interaction is not strong at elevated temperatures because dielectric constant inversely correlates with coulomb force. Ion pairs such as alkali and alkali earth chlorides (NaCl, KCl, $CaCl_2$, $MgCl_2$ etc.) and transition metal complexes with hard ligands (OH^-, Cl^-) are stable at elevated temperatures. In general number of ligands combining with cations decreases with increasing temperature.

1.10 Partitioning of Elements Between Aqueous Solution and Crystal

1.10.1 Partitioning of Element Between Aqueous Solution and Solid Solution Mineral

Ion exchange reaction is represented by

$$AM + B = BM + A \tag{1.89}$$

where *AM* and *BM* are *AM* and *BM* components in solid solution (*A,B*) *M*, respectively and *A* and *B* are dissolved ions.

In reaction (1.89) charges are omitted for a simplicity. Equilibrium constant of (1.89) is expressed as

$$K_{1\text{-}79} = (a_{BM}a_A)/(a_{AM}a_B) \tag{1.90}$$

$$= (\gamma_{BM}X_{BM}\gamma_{AM}m_A)/(\gamma_{AM}X_{AM}\gamma_{BM}m_B) \tag{1.91}$$

where a is activity, γ is activity coefficient, X is mole fraction, and m is molality.

1.10 Partitioning of Elements Between Aqueous Solution and Crystal

Distribution coefficient (Kd) is given by

$$Kd = (X_{BM}/X_{AM})/(m_B/m_A) \tag{1.92}$$

Therefore, the relation between distribution coefficient, Kd, and equilibrium constant, $K_{1\text{-}79}$ is represented by

$$K_{1\text{-}79} = \emptyset Kd \tag{1.93}$$

where $\emptyset = (\gamma_{BM}\gamma_A)/(\gamma_{AM}\gamma_B)$.

Dissolution reactions of AM and BM are

$$AM = A + M \tag{1.94}$$

$$BM = B + M \tag{1.95}$$

Equilibrium constants for reactions (1.94) and (1.95) are

$$K_{1\text{-}84} = a_A a_M/a_{AM} \text{ and } K_{1\text{-}85} = a_B a_M/a_{BM} \tag{1.96}$$

From (1.94), (1.95) and (1.96), we obtain

$$K_{1\text{-}84}/K_{1\text{-}85} = K_{1\text{-}79} = \emptyset Kd \tag{1.97}$$

Therefore

$$\log(K_{1\text{-}84}/K_{1\text{-}85}) = \log K_{1\text{-}79} = \log Kd + \log \emptyset \tag{1.98}$$

If $\emptyset = 1$ and distribution coefficient experimentally obtained is for equilibrium state, experimental distribution coefficient correlates positively with solubility product ratio with 1:1 slope. However, the experimental values deviate significantly from theoretical line derived here (Fig. 1.35).

The relationships between experimental distribution coefficient values for rhombohedral carbonates (Cd, Zn, Cu, Mn, etc.) are given by the Eqs. (1.99) and (1.100) (Rimstidt et al. 1998).

$$Kd = 1.6(K_{MCO_3}/K_{TrCO_3})^{0.57} \text{ (calcite)} \tag{1.99}$$

$$Kd = 4.1(K_{MCO_3}/K_{TrCO_3})^{0.57} \text{ (siderite)} \tag{1.100}$$

where Kd is distribution coefficient, K_{MCO_3} is solubility product of MCO_3 (M: Ca^{2+}, Fe^{2+}), and K_{TrCO_3} is solubility product of $TrCO_3$ (Tr: +2 cation).

Rimstidt et al. (1998) considered that experimental distribution coefficient values are determined by rate of formation of minerals from aqueous solution and diffusion rate of ion near the growing surface of mineral. Concentration profile of

Fig. 1.35 Plot of K_d (partition coefficient) for marine carbonates (Masuda 1986) on $\log K_d - \log K_{CaCO_3}/K_{TrCO_3}$ diagram (Rimstidt et al. 1998). *Open triangle*: marine carbonates, the other symbols: abiotic calcite

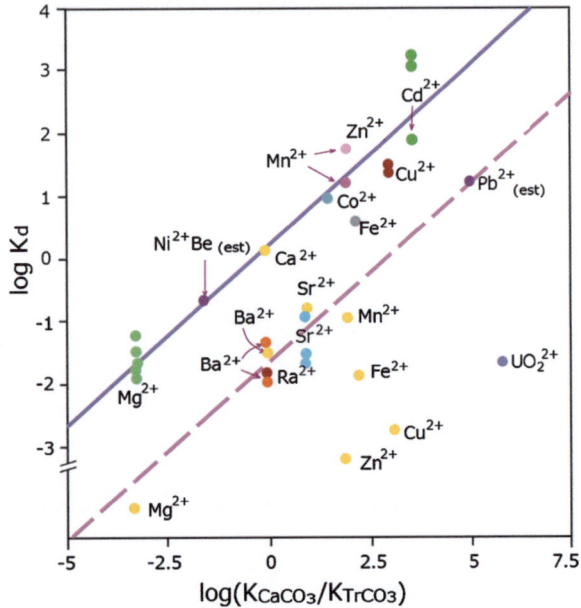

cations with the distance from the surface of crystal differs for the cases when Kd is more than 1 and less than 1.

Generally, concentration of elements in bulk solutions determined by the experiments is not that of the solution in contact with crystal surface. Therefore, experimentally determined Kd values deviate from equilibrium ones and are between equilibrium ones and 1. Rimstidt et al. (1998) indicated that the precipitation and diffusion rates are important factors controlling the distribution coefficients for minor elements between aqueous solution and carbonates and a given relationship between the distribution coefficients and solubility products for carbonates is obtained.

Next, influence of ø on Kd is considered below.

Gibbs free energy for exchange reaction is expressed as, $G_{ex} = G_{reaction} + G_{elastic}$

where $G_{reaction}$ is Gibbs free energy for reaction, and $G_{elastic}$ is Gibbs free energy for elasticity.

Nagasawa (1966) theoretically derived based on elastic model that log Kd linearly correlates with $-(r_o - r)^2/r_o^2$ (where r_o is ionic radius of major element, and r is ionic radius of trace element). The Nagasawa theory was applied to the partitioning of trace elements between magma and phenocryst and hydrothermal solution and mineral (Morgan and Wandless 1980; Blundy and Wood 1994; Wood and Blundy 2004).

Gnanapragasam and Lewis (1995) reasonably explained experimentally determined Kd for the partitioning of Ra between calcite and aqueous solution based on the equation

1.10 Partitioning of Elements Between Aqueous Solution and Crystal

$$-RTlnKd = 0.2G_{0ex} + 0.2G_{elastic} - 2.3 \qquad (1.101)$$

where G_{0ex} is Gibbs free energy for ion exchange reaction, and $G_{elastic}$ is Gibbs free energy for elasticity.

Gnanapragasam and Lewis (1995) estimated $G_{elastic}$ using the equation

$$G_{elastic} = 4\pi E N_0 m_t \\ + \left\{ (r_t - r_m)^2 / r_m^2 \right\} r_m^3 x \left\{ (1 + \mu) + 2(1 - 2\mu)(r_t/r_m)^3 \right\}^2 \right\} \qquad (1.102)$$

where E is Yang module, N_o is Avogadro number, m_t is mole fraction of trace element ion in solid solution, r_t is crystal ionic radius for trace element ion, r_m is crystal ionic radius for major element ion, and μ is Poisson ratio.

Equation (1.101) is converted into

$$-RTlnKd = -0.2RTlnK + 0.2G_{elastic} - 2.3 \qquad (1.103)$$

$$logKd = 0.2logK - (0.2G_{elastic} - 2.3)/2.3RT \qquad (1.104)$$

where K is equilibrium constant for the reaction, AM + B = BM + A, and is represented by

$$K = (a_{BM}a_A)/(a_M a_B) = (\gamma_M X_{BM} \gamma_A m_A)/(\gamma_{AM} X_{AM} \gamma_B m_B) \qquad (1.105)$$

Therefore,

$$logKd = 0.2(K_{spA}/K_{spB}) + 0.2log\phi - 0.085G_{elastic}/T + 1 \qquad (1.106)$$

Equation (1.106) means that a slope is 0.2 on logKd − log(K_{spA}/K_{spB}) diagram. However, the slope derived by Rimstidt et al. (1998) is 0.57 which is not consistent with that mentioned above. This inconsistency may be caused by the influence of the factors other than solubility product (e.g., ionic radii, Poisson ratio, Yang module etc.).

The relationships between Kd for marine carbonates and ionic radii are summarized in Fig. 1.36. These parabolic relationships can be explained by equations by Nagasawa (1966) and Gnanapragasam and Lewis (1995). Therefore, it is likely that the partitioning of trace elements between aqueous solution and minerals in previous experiments is governed by physical features (Yang module, Poisson ratio). In addition to the parameters mentioned above, the apparent distribution of elements between aqueous solution and mineral depends on chemical compositions of aqueous solution, ionic strength, pH, growth rate of mineral, degree of supersaturation etc. The more detailed discussion on the factors controlling the distribution coefficient is given in Shikazono and Ogawa (2007) and Prieto and Stoll (2010).

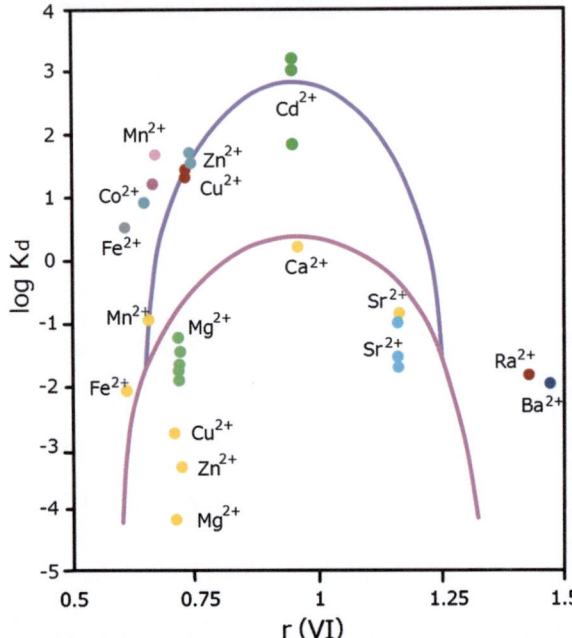

Fig. 1.36 Partition coefficient (K_d)—ionic radii (r(V1)) diagram for abiotic and biotic carbonates (Masuda 1986; Rimstidt et al. 1998). *Open triangle*: marine carbonates, the other symbols: abiotic carbonate

1.10.2 Rayleigh Fractionation

Above analysis assumes an attainment of equilibrium between aqueous solution and crystals. It is considered below that aqueous solution is in equilibrium with surface of precipitating crystal. This kind of process is called Rayleigh fractionation.

Let us consider precipitation of (Ca, Mn)CO$_3$ solid solution from aqueous solution. It is assumed that small amounts of precipitation, d[MnCO$_3$] and d[CaCO$_3$], are in equilibrium with aqueous solution. This equilibrium is expressed as

$$d[MnCO_3]/d[CaCO_3] = k m_{Mn^{2+}}/m_{Ca^{2+}} \qquad (1.107)$$

where k is Doener–Hoskins distribution coefficient (Doener and Hoskins 1925), $m_{Mn^{2+}}$ is Mn^{2+} concentration in aqueous solution, $m_{Ca^{2+}}$ is Ca^{2+} concentration in aqueous solution, and $\gamma_{Mn^{2+}}/\gamma_{Ca^{2+}}$ is assumed to be unity.

Integrating (1.107),

$$\int_i^f d[MnCO_3]/m_{Mn^{2+}} = \int_i^f d[CaCO_3]/m_{Ca^{2+}} \qquad (1.108)$$

where i is initial state, f is final state, and k is Doener–Hoskins distribution coefficient, we obtain

$$\log(m_{\text{Mni}^{2+}}/m_{\text{Mnf}^{2+}}) = k\log(m_{\text{Ca}^{2+}_i}/m_{\text{Ca}^{2+}_f}) \qquad (1.109)$$

This expression is known as the logarithmic distribution law (Doener–Hoskins rule). Chemical compositions of crystal and aqueous solution vary depending on k.

Diffusion rate of cations in crystal is very slow, resulting to heterogeneous composition in crystal. It has been experimentally obtained for many heterogeneous system ((Ba, Sr) SO_4–H_2O system, (Ca, Sr)SO_4–H_2O system etc.) that Doener–Hoskins rule is established (Doener and Hoskins 1925; Gordon et al. 1959; Shikazono and Holland 1983). However it is noted that true Doener–Hoskins behavior is expected to occur only when growth rates are slow enough to submit equilibrium between the solid-phase increments and solution fast enough to prevent recrystallization of the growth crystals (Glynn et al. 1990). For example, Shikazono and Holland (1983) studied the partitioning of Sr between anhydrite and mixtures of seawater with NaCl solution to temperatures between 150 and 250 °C. They found that the Doener–Hoskins type distribution coefficient in this system depends on less on the NaCl concentration, temperature, and precipitation rate than in the degree of supersaturation of the solution with respect to anhydrite and/or the morphology of the precipitated anhydrite crystals (acicular-type and rectangular type).

Compositional variations of minerals occurring in hydrothermal ore deposits (e.g., barite (Ba,Sr)SO_4, carbonate (Ca,Sr)CO_3)), metamorphic rocks (garnet, $X_3Y_2(SO_4)_3$, X = Ca^{2+}, Mg^{2+}, Fe^{2+}, Mn^{2+}, Y = Al^{3+}, Fe^{3+}, Cr^{3+}) and minerals crystallizing from magma are interpreted in terms of Rayleigh fractionation model.

Cited Literature

Ahrland S, Chatt J, Davies NR (1958) Quart Rev London 12:265–276
Anbeek C (1992) Geochim Cosmochim Acta 56:3957–3970
Anderson DE (1981) In: Reviews in mineralogy (Am Min), vol 8, pp 211–260
Arnórsson S, Sigurdsson S, Svarvarsson H (1982) Geochim Cosmochim Acta 46:1513–1532
Arnórsson S, Gunnlaugsson E, Svarvarsson H (1983) Geochim Cosmochim Acta 47:547–566
Barnes HL (1979) In: Barnes HL (ed) Geochemistry of hydrothermal ore deposits. Wiley, New York, pp 404–409
Barnes HL, Kullerud G (1961) Econ Geol 56:648–688
Barton PB Jr (1984) Fluid-mineral equilibria in hydrothermal system. In: Reviews in economic geology, vol 1, pp 99–114
Blundy J, Wood B (1994) Nature 372:452–454
Brimhall GH, Crerar DA (1987) In: Carmichael ISE, Eugster HP (eds) Thermodynamic modeling of geological materials-minerals, fluids, and melts. Reviews in Mineralogy (Am Min), vol 17, pp 235–322
Crerar DA (1975) Geochim Cosmochim Acta 39:1375–1385
Crerar DA, Barnes HL (1976) Econ Geol 71:772–794
Doener HA, Hoskins WM (1925) Chem Soc Am J 47:662–675
Fournier RO (1973) In: Proceedings of the international symposium on hydrogeochemistry and biogeochemistry, Japan (1970), vol 1. Hydrogeochemistry, Clard JW, Washington DC, pp 122–139

Fournier RO, Truesdell AHC (1973) Geochim Cosmochim Acta 37:1255–1275
Garrels RM (1967) In: Abelson PH (ed) Researches in geochemistry. Wiley, New York
Garrels RM, Mackenzie FT (1967) Origin of the chemical compositions of some springs and lakes. In: Gould RF (ed) Equilibrium concepts in natural waters, vol 67, Advances in chemistry Series. American Chemical Society, Washington DC, pp 222–242
Glynn PD, Reardon EJ (1990) Am J Sci 290:164–201
Glynn PD, Reardon EJ, Plummer LN, Busenberg E (1990) Geochim Cosmochim Acta 541:267–282
Gnanapragasam E, Lewis B (1995) Geochim Cosmochim Acta 59:5103–5111
Goldich SS (1938) J Geol 46:17–58
Harvie CE, Moller N, Weave JH (1984) Geochim Cosmochim Acta 48:723–751
Helgeson HC (1969) Am J Sci 267:729–804
Helgeson HC (1971) Geochim Cosmochim Acta 35:421–469
Helgeson HC (1972) Geochim Cosmochim Acta 36:1067–1070
Helgeson HC (1974) In: Mackenzie WS, Zussman J (eds) The feldspars. Manchester University Press, Manchester, pp 184–217
Helgeson HC (1979) In: Barnesw HL (ed) Geochemistry of hydrothermal ore deposits. Wiley, New York, pp 568–610
Helgeson HC, Kirkham DH (1974) Am J Sci 274:1199–1261
Helgeson HC, Kirkham DH, Flowers GC (1981) Am J Sci 281:1241–1516
Helgeson HC, Murphy WM, Aagaard P (1984) Geochim Cosmochim Acta 48:2405–2432
Hemley JJ, Jones WR (1964) Econ Geol 59:538–569
Henley RW (1984) In: Reviews in economic geology, vol 1, pp 115–128
Meyer C, Hemley JJ (1967) Wall rock alteration. In: Barnes HL (ed) Geochemistry of hydrothermal ore deposits. Holt Rinehart Winston, New York, pp 166–235
Morgan JW, Wandless GA (1980) Geochim Cosmochim Acta 44:973–980
Nagasawa H (1966) Science 152:767
Nesbitt HW, Macral ND, Shotyk W (1991) J Geol 99:429–442
Parkhurst DL, Thorstenson DC, Plummer C\N (1990) PHREEQE-a computer program for geochemical calculations. US Geological Survey Water Resources Investigations Report, pp 80–96
Pearson RG (1963) J Am Chem Soc 85:3533–3539
Pearson RG (1968) J Chem Educ 45:581–587, 643–648
Plummer LN, Prestemon EC, Parkhurst DL (1991) An interactive code (NETPATH) for modeling NET geochemical reactions along a flow PATH Version 2.0 U S Geol Survey Water Resources Invest, pp 94–4169
Rimstidt JD, Balog A, Webb J (1998) Geochim Cosmochim Acta 62:1851–1863
Rose A, Burt DM (1979) Hydrothermal alteration. In: Barnes HL (ed) Geochemistry of hydrothermal ore deposits. Wiley, New York, pp 173–235
Shikazono N (1976) Geochem J 10:47–50
Shikazono N (1978a) Econ Geol 73:524–533
Shikazono N (1978b) Chem Geol 23:239–254
Shikazono N (1985) Chem Geol 49:213–230
Shikazono N (1988) Min Geol Spec Issue 12:47–55
Shikazono N (2002) Jpn Mag Min Petrol Sci 31:197–207 (in Japanese with English abstract)
Shikazono N, Holland HD (1983) Econ Geol Mon 5:320–328
Shikazono N, Ogawa Y (2007) Genshiryoku Back End Kenkyu 12:3–9 (in Japanese with English abstract)
Shikazono N, Shimizu M (1992) Can Min 30:137–143
Shikazono N, Holland HD, Quirk RF (1983) Econ Geol Mon 5:329–344
Susak N, Crerar D (1982) Econ Geol 77:476–482
Susak N, Crerar D (1985) Geochim Cosmochim Acta 49:555–564
Thompson JB Jr (1967) In: Abelson PH (ed) Researches in geochemistry, vol 2. Cummings Publishing Co Inc, Menlo Park, pp 129–142

Uchida E, Goryozono Y, Naito M (1996) Geochem J 30:99–109
Wagman DD, Evans WH, Parker VB, Schumm RH, Halow I, Bailey SM, Chuney KL, Nuttall RL (1982) J Phys Chem Ref Data 11(2):392
Wood BJ, Blundy JD (2004) Treatise on geochemistry, vol 2, The mantle and core. Elsevier, Oxford, pp 395–424
Yamada S, Tanaka M (1975) Inorg Nucl Chem 37:587–589
Yamakawa M (1991) The third symposium on advanced nuclear energy research—Global environment and nuclear energy, pp 150–158

Further Reading

Barnes HL (ed) (1967) Geochemistry of hydrothermal ore deposits. Rinehart and Winston, Inc, Holt
Barnes HL (ed) (1979) Geochemistry of hydrothermal ore deposits. Wiley, New York
Davies CW (1962) Ion association. Butterworth, Washington, DC
Garrels RM, Christ CL (1965) Solutions, minerals, and equilibria. Harper & Row/John Weatherhill, Inc, New York/Tokyo
Gordon L, Salutsky MC, Willard HH (1959) Precipitation from homogeneous solution. Wiley, New York
Guggenheim EA (1952) Mixtures. Oxford University Press, London
Guggenheim EA (1967) Thermodynamics, 5th edn. North-Holland Publishing Co., Amsterdam, p 390
Helgeson HC (1979) Mass transfer among minerals and hydrothermal solutions. In: Barnes HL (ed) Geochemistry of hydrothermal ore deposits, 2nd edn. Wiley, New York, pp 568–631
Marini L (2007) Geological sequestration of carbon dioxide. Elsevier, Amsterdam
Nordstrom DK, Munoz JL (1985) Geochemical thermodynamics. The Benjamin/Cummings Publishing Co. Inc, Menlo Park
Pitzer KS (ed) (1991) Activity coefficients in electrolyte solutions, 2nd edn. CRC, Boca Raton, p 542
Prieto M, Stoll H (2010) Ion partitioning in ambient-temperature aqueous system, vol 10, EMU in mineralogy. The Mineralogical Society of Great Britains & Ireland, London
Prigogine I (1967) Introduction to thermodynamics of irreversible processes. Wiley, New York
Shikazono N (2003) Geochemical and tectonic evolution of arc-backarc hydrothermal systems: implication for the origin of Kuroko and epithermal vein-type mineralization and the global geochemical cycle, vol 8, Developments in geochemistry. Elsevier, Amsterdam
Shikazono N (2010) Introduction to earth and planetary system science. Springer
Stumm W, Morgan JJ (1970) Aquatic chemistry. Wiley, New York
Stumm W, Morgan JJ (1981) Aquatic chemistry, 2nd edn. Wiley, New York
Stumm W, Morgan JJ (1996) Aquatic chemistry, 3rd edn. Wiley, New York

Chapter 2
Partial Chemical Equilibrium

The concept of partial and local chemical equilibrium in natural system has been proposed and developed by Thompson (1959) and Helgeson (1979). After that many application and computer simulations based on partial chemical equilibrium have been conducted (e.g., Wolery 1983; Reed 1983). The principle of this model will be briefly described and application of this model to hydrothermal system accompanied by hydrothermal alteration and formation of hydrothermal ore deposits will be given below.

2.1 Water–Rock Interaction

Magma intrudes continuously at mid-oceanic ridge, causing high heat flow rate. Seawater penetrating into deep underground along fractures and cracks becomes to hydrothermal solution interacting with oceanic crust. Mid-oceanic ridge basalt is suffered hydrothermal alteration and metamorphism. Actinolite ($Ca_2(Mg,Fe)_5$, $Si_8O_{22}(OH)_2$) and epidote ($Ca_2FeAl_2Si_3O_{12}(OH)$) alterations are observed in basalt near mid-oceanic ridge axis. The basaltic rocks away from the axis are suffered chlorite and smectite alterations. The alteration process of seawater–basalt reaction is interpreted in terms of partial chemical equilibrium model described below.

Let us consider the interaction of small amounts of fresh rocks with aqueous solution. These rocks and aqueous solution are not in equilibrium with each other. These rocks and aqueous solution react to reach to equilibrium state at constant temperature and pressure. Alteration minerals form and chemical compositions of aqueous solution change during this initial step.

Next, if we put small amounts of fresh rocks to this system, alteration of rocks and the change in chemical composition of aqueous solution occur, resulting to equilibrium state. The progress variable for the reaction corresponds to water/rock ratio. Different alteration occurs, depending on chemical and mineralogical compositions of initial aqueous solution and rock, temperature and pressure.

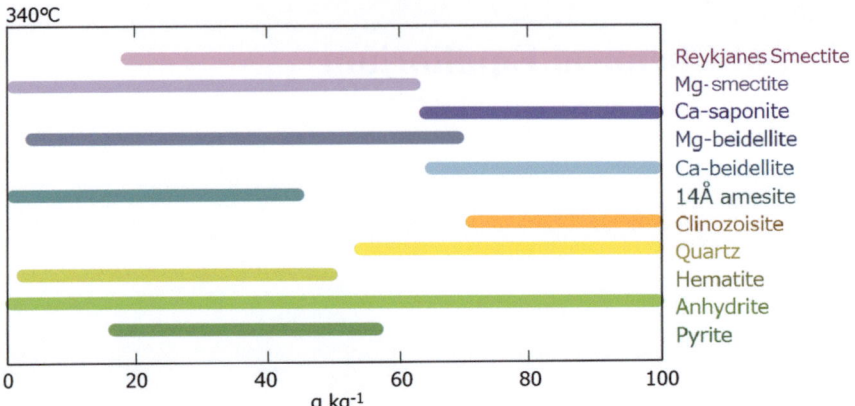

Fig. 2.1 The change in alteration minerals formed during seawater–basalt reaction (Wolery 1978)

The progress variable for the reaction is defined below (Prigogine and Defay 1954; Helgeson 1979).

The following reaction is considered.

$$\nu_1 R_1 + \nu_2 R_2 + \nu_3 R_3 + \ldots + \nu_j R_j \rightarrow \nu_{j+1} P_{j+1} + \nu_{j+2} P_{j+2} \ldots + \nu_e P_e \qquad (2.1)$$

where $R_1, \ldots R_j$ are reaction substance, $P_{j+1}, \ldots P_e$ is product, $\nu_1, \ldots \nu_e$ is mol coefficient for component 1…., e or chemical stoichiometric coefficient.

Increment of mass is proportional to M_i and ν_i. Thus

$$\left.\begin{aligned} m_1 - m_{10} &= \nu_1 M_1 \varsigma \\ \ldots \quad \ldots &\quad \ldots\ldots\ldots \\ \ldots \quad \ldots &\quad \ldots\ldots\ldots \\ m_i - m_{io} &= \nu_i M_i \varsigma \\ m_c - m_{co} &= \nu_c M_c \varsigma \end{aligned}\right\} \qquad (2.2)$$

where ς is progress variable for the reaction.

The result of calculation on the change in the alteration minerals formed during seawater–basalt reaction at elevated temperatures is shown in Fig. 2.1 (Wolery 1978; Reed 1983) which corresponds closely to that of experimental studies of seawater–basalt reaction (e.g., Bischoff and Dickson 1975) and natural greenstone from ophiolite, seasloor and Tertialy volcanic rock mining area (e.g., Shikazono et al. 1995). For instance, SO_4^{2-} and Mg concentrations decrease by the reaction (Fig. 2.2), while the concentrations of the other elements (Si, base metals, Ca etc.) increase. The decrease in Mg concentration in hydrothermal solution is caused by the formation of Mg-minerals (e.g., Mg-chlorite, Mg-smectite). SO_4^{2-} concentration decreases with increasing temperature which is due to the precipitation of anhydrite ($CaSO_4$). Anhydrite precipitates from heated seawater because its solubility decreases with increasing temperature (Holland and Malinin 1979).

Fig. 2.2 The concentration of SO_4^{2-}, dissolved SiO_2, Na^+ and Cl^- in Copenhagen seawater during interaction with basalt powder at 200 °C and 500 bar (Bischoff and Dickson 1975)

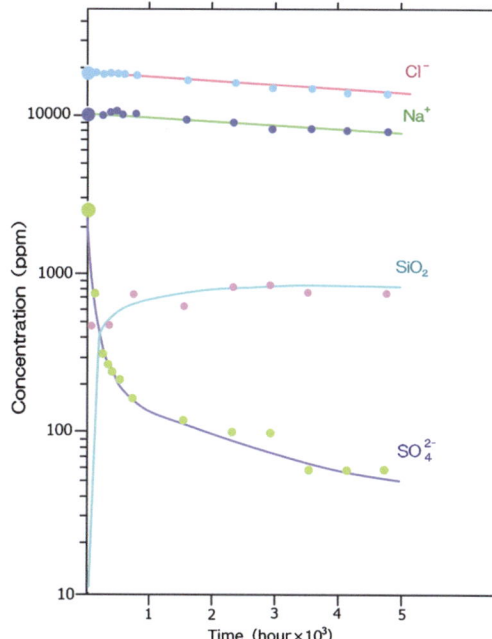

Chemical and mineralogical compositions of altered basalt depend on water/rock ratio (Mottl 1983; Shikazono et al. 1995). For example, epidote forms at low water/rock ratio W/R (<1) (Shikazono 1984), while chlorite is stable at high W/R (1–100) (Shikazono and Kawahata 1987).

Field observation and analytical studies on altered rocks are consistent with those of computer simulations on basalt–seawater interaction at elevated temperatures (Mottl 1983; Shikazono et al. 1995).

2.2 Hydrothermal Alteration Process in Active and Fossil Geothermal Areas

The rocks in active and fossil geothermal areas are suffered hydrothermal alterations and the causes for hydrothermal alteration zoning are considered in terms of partial chemical equilibrium model (Giggenbach 1981; Shikazono 1988; Takeno 1989).

Let us consider the compositional change of ascending hydrothermal solution of meteoric water origin. Alkali earth and alkali elements dissolve to the solution by the interaction of relatively low temperature solution. At higher temperatures, hydrothermal solution becomes to be in equilibrium with Na-feldspar and K-feldspar. The equilibrium constant of the reaction, K-feldspar + Na^+ = Na-feldspar + K^+, $K = m_{K^+}/m_{Na^+}$, where $a_{K\text{-}f}$, $a_{Na\text{-}f}$ and $\gamma_{Na^+}/\gamma_{K^+}$ ratio are

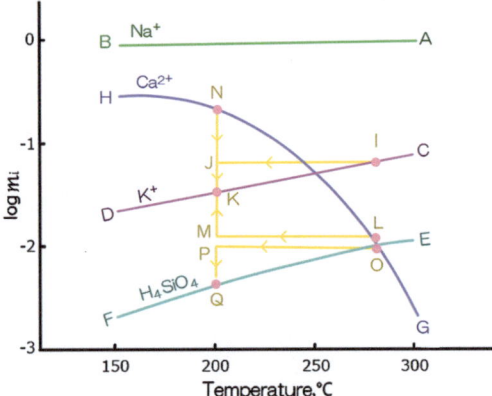

Fig. 2.3 The dependence of concentration of K^+, Na^+, Ca^{2+} and H_4SiO_4 in equilibrium with common alteration minerals (K-feldspar, Na-feldspar, quartz) on temperature and the changes in the concentrations of hydrothermal solution (Na^+, Ca^{2+}, K^+, H_4SiO_4) during the ascending of hydrothermal solution and precipitation of alteration on minerals (Shikazono 1988). Thermochemical data used for the calculations are from Helgeson (Helgeson 1979). Calculation method is given in Shikazono (1978). Chloride concentration in hydrothermal solution is assumed to be 1 mol/kg · H_2O. A, B: Na^+ concentration in solution in equilibrium with low albite and adularia. C, D: K^+ concentration in solution in equilibrium with low albite and adularia. E, F: H_4SiO_4 concentration in solution in equilibrium with quartz. G, H: Ca^{2+} concentration in solution in equilibrium with low albite and adularia

assumed to be unity, increases with increasing temperature (Fig. 2.3). It is considered that the hydrothermal solution initially in equilibrium with potassic alteration minerals ascends rapidly and interacts with country rocks at lower temperatures (I → J → K in Fig. 2.3). In this case, addition of K^+ to the rocks takes place. K-bearing minerals such as sericite and K-feldspar precipitate from the fluid accompanied by the destruction of plagioclase in the country rocks and liberation of Ca, Sr and Na to the fluids. Dissolutions of plagioclase occur by this mechanism. It is expected that SiO_2 content of the country rocks increases with progressive alteration because solubility of SiO_2 decreases with a decrease in temperature (O → P → Q in Fig. 2.3).

Na-feldspar, Ca-minerals (anhydrite, analcime, Ca-zeolites) and Mg-minerals (e.g., Mg-chlorite, Mg-smectite) form at recharge zone, while K-feldspar and K-mica at discharge zone. This is caused by the adiabatic ascending of hydrothermal solution initially in equilibrium with K-feldspar and Na-feldspar without the interaction with surrounding rocks. The enrichment of SiO_2 at discharge zone is related to the increasing SiO_2 solubility with increasing temperature. SiO_2 minerals (quartz, opal, chalcedony, α-cristobalite) precipitate with decreasing temperature. The distribution of the other minerals in geothermal system can be explained in such thermochemical calculations.

2.3 Oxygen Isotopic Variations During Water–Rock Reaction, Mixing and Boiling of Fluids

The following equations can be used to derive the relationship between $\delta^{18}O$ of rocks (minerals) and water/rock ratio for oxygen isotopic variation during water–rock reaction in a closed system.

$$W_i \delta^{18}O_{iW} + R_i \delta^{18}O_{iR} = W_f \delta^{18}O_{fW} + R_f \delta^{18}O_{fR} \tag{2.3}$$

where W_i is mass of oxygen in initial water, R_i is mass of oxygen in initial rock, W_f is mass of oxygen in final water, R_f is mass of oxygen in final rock, i is initial state, and f is final state.

It is approximated as, $W_i = W_f$ and $R_i = R_f$. Thus, they are expressed as W and R, respectively, below.

Generally, feldspar is dominant silicate mineral in rocks. Thus, it is assumed that

$$\delta^{18}O_{fR} = \delta^{18}O_{feldspar} \tag{2.4}$$

Oxygen isotopic exchange equilibrium between feldspar and water is assumed, providing the following equation.

$$\delta^{18}O_W = \delta^{18}O_{feldspar} - \Delta \tag{2.5}$$

where $\Delta = 10^3 \ln\alpha$, and α is oxygen isotopic fractionation factor between feldspar and water that is represented as a function of temperature.

Calcite is common alteration mineral. $\delta^{18}O$ of calcite is represented by

$$\delta^{18}O_{calcite} = \delta^{18}O_{feldspar} - \Delta \tag{2.6}$$

where $\Delta' = 10^3 \ln\alpha'$, and α' is oxygen isotopic fractionation factor between feldspar and calcite.

Above equations give $\delta^{18}O_{calcite}$ as functions of water/rock ratio (W/R), $\delta^{18}O_{iw}$, and $\delta^{18}O_{fR}$ (Fig. 2.4).

Above calculations were made for a closed system (Shikazono 1989). Similar calculations can be done for an open system (Taylor 1997).

$\delta^{18}O$ of alteration minerals such as calcite and hydrothermal solution change not only by water–rock interaction, but also by the mixing and boiling of fluids (Shikazono 1989).

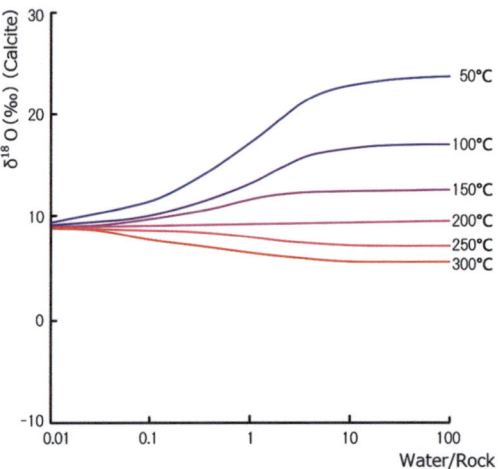

Fig. 2.4 Relationship between $\delta^{18}O$ of calcite and water/rock ratio (Shikazono 1988). Initial $\delta^{18}O$ of water and rock prior to the water–rock interaction is assumed to be 0 ‰ and +8 ‰, respectively

2.4 Formation of Minerals Accompanied by Separation of Vapor Phase and Liquid Phase and Boiling

The values of physicochemical variables (temperature, pH etc.) of aqueous solution change when it boils and vapor forms. This change causes the formation of minerals. The precipitated amounts of minerals due to boiling of hydrothermal solution in an open system can be calculated, based on partial equilibrium model (Reed and Spycher 1984, 1985; Drummond and Ohmoto 1985). In order to perform this type of calculation, thermochemical data on gaseous and dissolved aqueous species and minerals and solubility of minerals are necessary.

CO_2 gas releases from the system when hydrothermal solution boils, resulting to an increase in pH. This pH change is obvious from the following reactions.

$$H_2CO_3 = H_2O + CO_2 \tag{2.7}$$

$$H_2CO_3 = H^+ + HCO_3^- \tag{2.8}$$

$$HCO_3^- = H^+ + CO_3^{2-} \tag{2.9}$$

The activity ratios, a_{K^+}/a_{H^+} and a_{Na^+}/a_{H^+}, increase with an increase in pH due to release of CO_2 gas from aqueous solution. If initial hydrothermal solution is in equilibrium with K-feldspar, K-mica, quartz and Na-feldspar which are common minerals in rocks and CO_2 releases from the system, a_{K^+}/a_{H^+} and a_{Na^+}/a_{H^+} increase and chemical composition of hydrothermal solution plots in a stability field of K-feldspar (Fig. 2.5). Thus, it is possible that K-feldspar forms from boiling solution. pH, H_2S and H_2 which release from the system change with boiling of fluid. Temperature decreases accompanied by the boiling. However, the change in temperature is not be considered in the case of K-feldspar formation.

Fig. 2.5 Stability field of Ca–K minerals; 260 °C (Ellis and Mahon 1977). *W* Wairakei (New Zealand), *B* Broadlands (New Zealand). *Arrows* indicate the changes in concentrations after degassing of CO$_2$ from geothermal waters

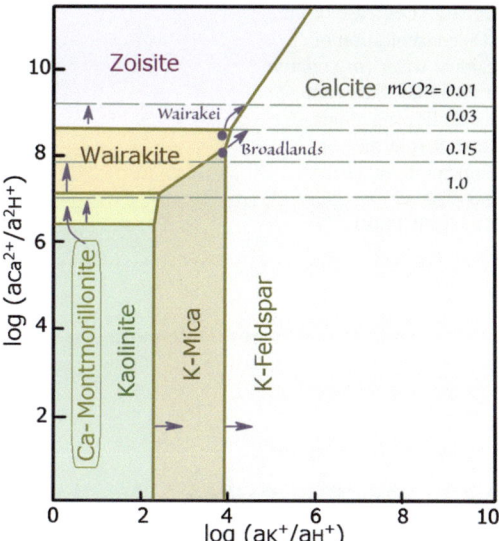

Fig. 2.6 Iso-solubility lines for barite and variations in (m$_{Na}$ + m$_K$) and temperature due to boiling and mixing (Hayba et al. 1985)

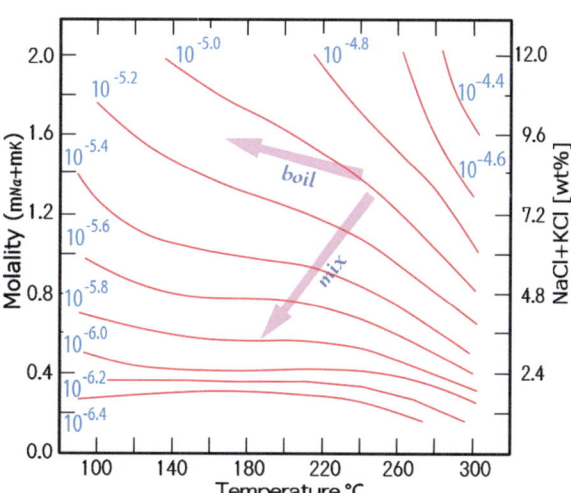

The influences of decreasing temperature due to boiling and mixing of hydrothermal solution with ground water on solubility of barite are shown in Fig. 2.6. The concentration of ionic species (Na$^+$, K$^+$, Ba^{2+} etc.) increases due to boiling. The trend in this decrease and decrease in temperature due to boiling cause precipitation of barite (Fig. 2.5). The mixing of hydrothermal solution with ground water causes efficient precipitation of barite. The precipitation mechanism is different for different type of boiling including one step, multi-step, isoenthalpy and Reighleigh boiling (Henley 1984). This will be considered below.

Fig. 2.7 Decrease of the CO_2 concentration of residual water (m_l) relative to the initial CO_2 content (m_o) following single-step steam loss in the temperature range 260–220 °C (Truesdell 1989)

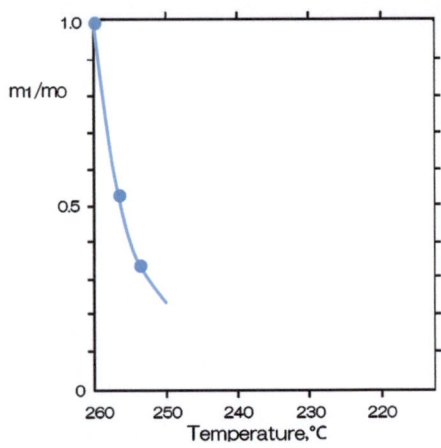

2.4.1 One Step Boiling

Mass balance relation for the adiabatic separation of CO_2 from deep geothermal water with m_o as initial CO_2 concentration in a closed system is expressed as (Truesdell 1984)

$$m_o = m_l(1-y) + m_v(y) \qquad (2.10)$$

where l is liquid phase, v is vapor phase, and y is fraction of vapor phase after the separation of CO_2.

The change in CO_2 concentration in liquid phase from 260 °C is shown in Fig. 2.7.

2.4.2 Multi-Step Boiling

Multi-step boiling causing fractionation of vapor phase from liquid phase is expressed as (Truesdell 1984)

$$m_l/m_o = [1/\{1 + y(B' - 1)\}]^n \qquad (2.11)$$

where n is number of step, and B' is average distribution coefficient for one-step boiling with a decrease in temperature.

This type of boiling is commonly observed in active and fossil geothermal systems. Evidence of boiling in fossil geothermal system is found in fluid inclusions in minerals.

Fluid inclusions are fluids trapped in crystal when it grows from fluids. Fluid inclusions generally consist of liquid phase and vapor phase. Usually they are dominantly composed of water. Aqueous solution contains cations (Na^+, K^+, Ca^{2+}, Mg^{2+}) and anions (Cl^-, HCO_3^-, SO_4^{2-}). Solid phases (NaCl, KCl etc.) and CO_2

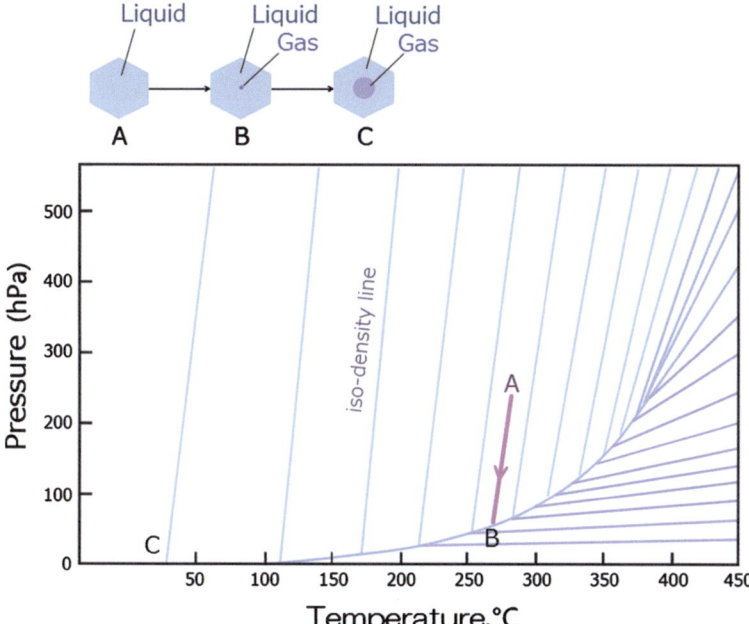

Fig. 2.8 Temperature and pressure variation of fluid inclusion (Shikazono 1988). Fluid inclusion formed at A changes as A → B → C

liquid sometimes exist in fluid inclusions. The vapor/liquid volume ratio in fluid inclusions in a crystal is widely variable when crystal forms from boiling fluid. Temperature when vapor and liquid phases homogenize by the heating is called homogenization temperature (or filling temperature). Homogenization temperature gives us useful information on temperature of formation of minerals. However we have to be careful about an estimate of formation temperature from homogenization temperature because the homogenization temperature is minimum temperature and is not true temperature of formation (Fig. 2.8).

2.5 Precipitation of Minerals Due to Mixing of Fluids

Several types of mixing of fluids (e.g., hydrothermal solution-seawater mixing, ground water-ground water mixing) near surface environment are occurring.

Chemical changes during the hydrothermal solution-seawater mixing associated with sekko ore in Kuroko deposits are considered below based on partial equilibrium model.

The sekko ore body dominately composed of anhydrite and gypsum occurs at lower horizon than sulfide horizons in Kuroko deposits* (Fig. 2.9). Sakai et al. (1970) indicated based on $\delta^{18}O$ and δD of gypsum ($CaSO_4 \cdot 2H_2O$)

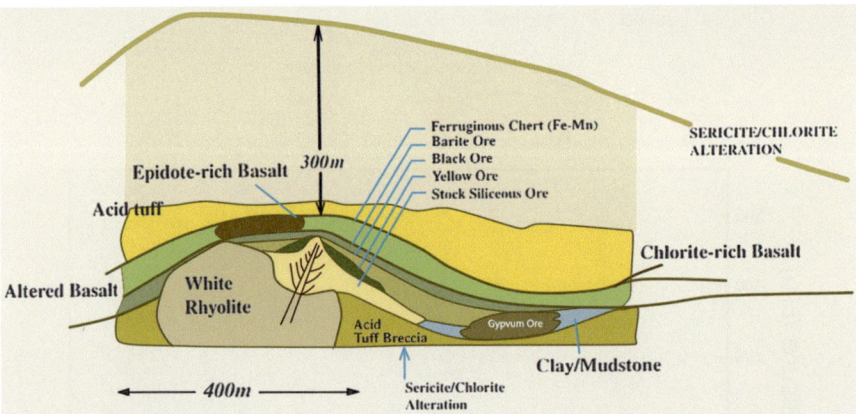

Fig. 2.9 Schematic distribution of Kuroko ore body and hydrothermally altered rocks (Sato 1974)

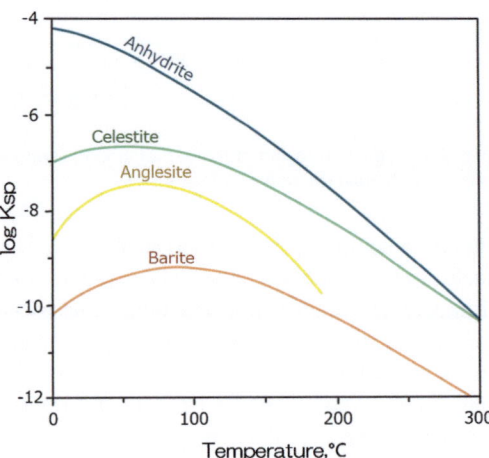

Fig. 2.10 The solubility products of several gangue sulfate minerals as a function of temperature (Naumov et al. 1974)

that gypsum was formed by the hydration of anhydrite ($CaSO_4$) at post-depositional stage. Solubility product (K_{sp}) of anhydrite decreases with increasing temperature (Fig. 2.10). If seawater is simply heated, anhydrite begins to precipitate at ca. 120 °C and 1 atm. It has been thought that anhydrite formed by the simple heating of seawater by dacite dome on seafloor from the $\delta^{34}S$ value of anhydrite (+21 ‰) that is nearly same to that of seawater and inverse relationship between solubility of anhydrite and temperature (Holland and Malinin 1979). However, if anhydrite precipitates due to this simple heating, Sr content of anhydrite calculated based on partition coefficient of Sr between anhydrite and hydrothermal solution ($=(X_{Sr}/X_{Ca})_{anhydrite}/(m_{Sr}/m_{Ca})_{seawater}$ where X is mole fraction and m is concentration) experimentally determined (Shikazono and Holland 1983a, b) is higher than the analytical data on Sr content of anhydrite from Kuroko deposits.

2.5 Precipitation of Minerals Due to Mixing of Fluids

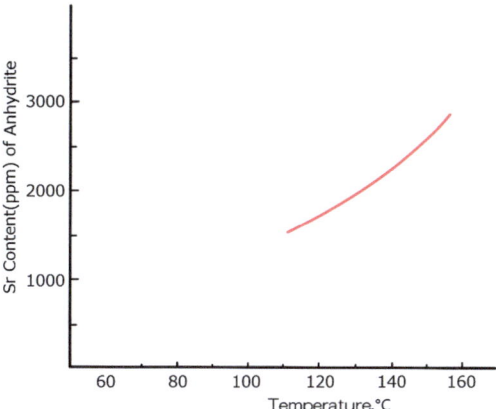

Fig. 2.11 Change in the strontium content of anhydrite precipitated during the heating of normal seawater without any seawater–rock interaction (Shikazono 1983)

The variation of Sr content of anhydrite with temperature during the simple heating of seawater can be obtained from the following equation (Shikazono et al. 1983).

$$K_{spo} = (m_{Ca^{0^{2+}}} - \Delta_0)(m_{SO_4^{2-}0} - \Delta_0) \tag{2.12}$$

where K_{spo} is solubility product, and Δ_0 is mole of anhydrite precipitated from the heated seawater per 1 kg H_2O during small increment of temperature.

At a given temperature, K_{sp0} is constant value.

If we take seawater values for m_{Cao} and $m_{SO_4^{2-}0}$, we can obtain Δ_0.

In next step, the solution is heated to a given temperature. At this temperature, we obtain the precipitated amount of anhydrite from the reactions.

$$m_{Ca1^{2+}} = m_{Ca0^{2+}} - \Delta_0 \tag{2.13}$$

$$m_{SO_4^{2-}1} = m_{SO_4^{2-}0} - \Delta_0 \tag{2.14}$$

$$K_{sp1} = (m_{Ca^{2+}1} - \Delta_1)(m_{SO_4^{2-}1} - \Delta_1) \tag{2.15}$$

Step-wise calculation on the precipitated amount of anhydrite during the heating of seawater gives the changes in Ca^{2+} and SO_4^{2-} concentrations.

From the result of calculation and partition coefficient of Sr between anhydrite and aqueous solution, we obtain

$$K_i = (m_{Ca^{2+}1}/m_{Sr^{2+}i})/(X_{Ca}/X_{Sr}) = (m_{Ca^{2+}1} \ m_{Sr^{2+}1})/\{(1-X_{Sr})/X_{Sr}\} \tag{2.16}$$

where X_{Ca} is mole fraction of $CaSO_4$ in anhydrite, and X_{Sr} is mole fraction of $SrSO_4$ in anhydrite.

The calculated result of the dependency of Sr content of anhydrite on temperature is shown in Fig. 2.11. This relation cannot explain Sr content of anhydrite from Kuroko deposits.

Next we consider the mixing of seawater and hydrothermal solution as a possible cause for the deposition of anhydrite. It is found that Sr content and $^{87}Sr/^{86}Sr$

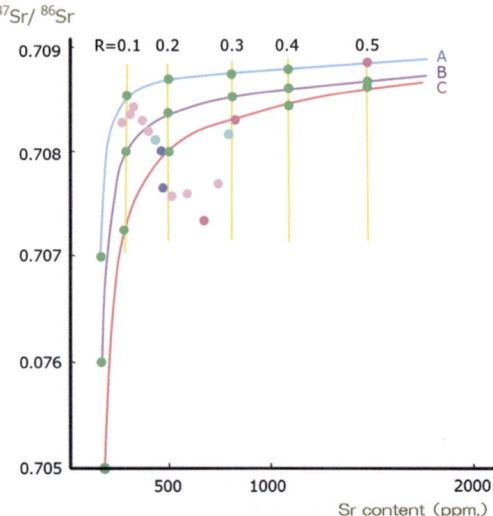

Fig. 2.12 Change of the strontium content and $^{87}Sr/^{86}Sr$ ratio of Kuroko anhydrite during the deposition and dissolution due to the mixing of hot ascending solution and cold solution (normal seawater) (Shikazono et al. 1983). R (mixing ratio (in weight)) = S.W./(S.W. + H.S.) in which S.W. and H.S. are seawater and hydrothermal solution, respectively. *Open triangle*: Fukazawa deposits, *Solid triangle*: Hanawa deposits, *Open square*: Wanibuchi deposits, *Solid square*: Shakanai deposits. Concentration of Ca^{2+}, Sr^{2+} and SO_4^{2-} of H.S. are assumed to be 1,000 ppm, 1 ppm, and 10^{-4} mol/kg · H_2O, respectively. Concentration of Ca^{2+}, Sr^{2+} and SO_4^{2-} of S.W. are taken to be 412, 8, and 2,712 ppm. Temperatures of H.S. and S.W. are assumed to be 350 and 5 °C (Shikazono et al. 1983)

of anhydrite can be explained by the mixing of seawater and hydrothermal solution (Fig. 2.12).

*Kuroko deposits: Kuroko deposits are stratabound-polymetallic (Cu, Pb, Zn, Au, Ag etc.) sulfide–sulfate deposits occurring in middle Miocene Green tuff region in Japan. The deposits are characterized by ore zoning from lower to upper stratigraphic horizons, Sekko (gypsum ore) zone → Keiko (siliceous ore) zone → Oko (yellow ore) zone → Kuroko (black ore) zone. Characteristic minerals and elements enriched in each ore zone are as follows.

Sekko zone (ore): anhydrite, gypsum (Appendix, Plate 11); Ca
Keiko (siliceous) zone (ore): quartz, chalcopyrite (Appendix, Plate 12), pyrite (Appendix, Plate 13); Si, Cu, Fe
Oko (yellow) zone (ore) (Appendix, Plate 14): chalcopyrite, pyrite; Cu, Fe
Kuroko (black) zone (ore): chalcopyrite, sphalerite (Appendix, Plate 15), galena (Appendix, Plate 16), pyrite, barite (Appendix, Plate 17); Zn, Pb, Fe, Cu, Ba, Au, Ag, As, Sb
Barite zone: barite, quartz; Ba, Si
Tetsusekiei (ferruginous chert) zone: hematite, quartz; Fe, Si

Detailed descriptions on Kuroko deposits can be referred in Shikazono (2003).

2.6 Formation of Hydrothermal Ore Deposits by Hydrothermal Solution-Seawater Mixing

2.6.1 Mid-Oceanic Ridge Deposits

It has been found that hydrothermal solution issues from the mid-oceanic ridges and hydrothermal ore deposits are forming on the seafloor and under the subseafloor environment. Initial chemical composition of hydrothermal solution prior to the mixing with cold seawater is estimated by extrapolating the analytical data to zero concentration of Mg (Fig. 2.13). Experimental data and thermochemical calculations on basalt–seawater interaction at elevated temperatures indicate that the hydrothermal solution after the interaction contains almost no Mg. The precipitated amounts of minerals during the mixing can be calculated (Fig. 2.14). At early and

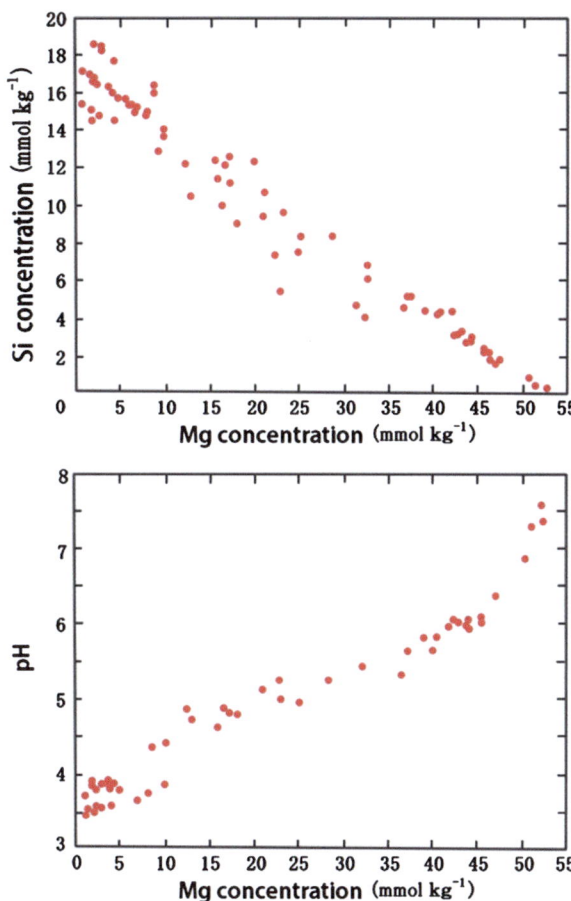

Fig. 2.13 Mg and Si concentrations and pH in ridge hydrothermal solution (Von Damm et al. 1985)

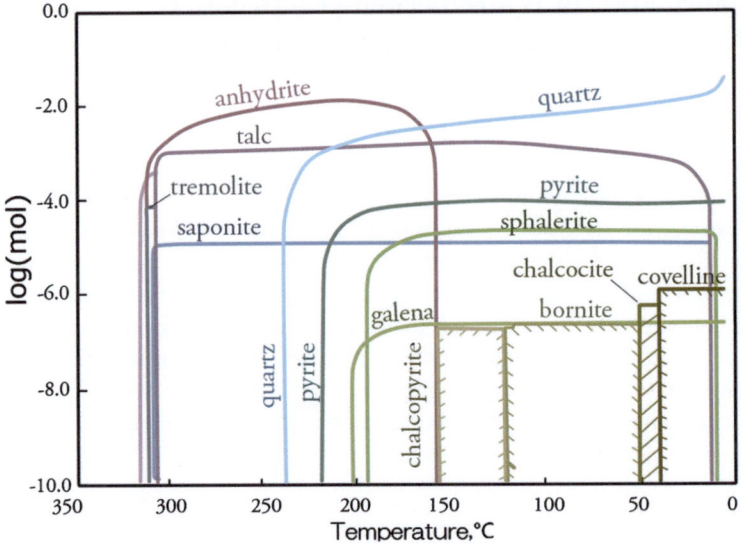

Fig. 2.14 Precipitated amounts of minerals per 1 kg mixed solution during mixing of hydrothermal solution (315 °C, 500 × 10^5 Pa) and seawater (2 °C) (Janecky and Shanks 1988)

high temperature stage of hydrothermal activity, chalcopyrite precipitates and at late and low temperature stage, anhydrite precipitates but it dissolves by cold seawater at waning stage of hydrothermal activity.

2.6.2 Kuroko Deposits

No hydrothermal solution (except fluid inclusions in minerals) is preserved in the ores of Kuroko deposits in Japan because they formed at past (ca.15 million years ago) and hydrothermal solution responsible for the ore deposits moved away from the depositional site. The calculations on the precipitation of minerals due to the mixing of seawater and hydrothermal solution is useful to know the precipitation sequence and temperature of mineralization.

Assuming that initial hydrothermal solution prior to the mixing with seawater is in equilibrium with ore-forming minerals, we can calculate the relationship between precipitated amounts of minerals and temperature (= mixing ratio). In general, the calculated results are not in agreement with the observation of the distribution of the minerals in ore deposits. For example, ore zonings of chimney and Kuroko ore body are not consistent with the results of calculations. Sulfate minerals ($BaSO_4$, $CaSO_4$) are abundant at the outer part of the chimney and sulfides tend to occur at inner parts, indicating earlier stage of sulfates precipitation than sulfides. In early stage of mixing both sulfides ($CuFeS_2$, ZnS, FeS_2) and sulfates ($CaSO_4$, $BaSO_4$) precipitate. The zoning of minerals in chimney will be interpreted in terms of precipitation kinetics-fluid flow model in Sect. 4.1.

2.7 Formation of Gold Deposits by Mixing of Hydrothermal Solution and Acid Ground Water

Gold (Appendix, Plate 18) is frequently enriched in active geothermal system and epithermal gold deposits*(Appendix, Plates 19–23).

*Epithermal deposits: The hydrothermal deposits which formed at relatively shallow environment from surface and at lower temperatures (less than 300 °C) compared with the other hydrothermal ore deposits are classed as epithermal deposits.

Solubility of gold in hydrothermal solution is important factor when we consider the depositional mechanism of gold in hydrothermal system.

If gold-thio-complex such as $Ag(HS)_2^-$ is dominant gold species in hydrothermal solution, gold solubility is controlled by the reaction.

$$Au + 2H_2S + 1/4O_2 = Au(HS)_2^- + H^+ + 1/2H_2O \quad (2.17)$$

Equilibrium constant for (2.16) is written as

$$K_{2-16} = \left(a_{Au(HS)_2^-}\, a_{H^+}\right) / \left(a_{H_2S^2} f_{O_2}^{1/4} a_{Au}\right) \quad (2.18)$$

Assuming $\sum Au$ (total dissolved gold concentration) $= m_{Au(HS)_2^-}$ and $a_{Au} = 1$, iso-gold concentration contours can be drawn on $\log f_{O_2}$-pH diagram at a given temperature, $\sum S$ and ionic strength (Henley 1984) (Fig. 2.15). It is found in Fig. 2.15 that gold concentration is high in neutral conditions and near SO_4^{2-}/H_2S boundary.

Gold is usually present as gold-silver solid solution (electrum) in the ore deposits. Therefore, the solubility of silver affects the solubility of gold. The reactions involving silver have to be considered. They are

$$Ag + 2H_2S + 1/4O_2 = Ag(HS)_2^- + H^+ + 1/2H_2O \quad (2.19)$$

$$Ag + 2HCl + 1/4O_2 = AgCl_2^- + 1/2H_2O + H^+ \quad (2.20)$$

From the equilibrium constants for (2.19) and (2.20), we can derive gold solubility as functions of pH, f_{O_2}, a_{H_2S}, $a_{S^{2-}}$, a_{Ag} and temperature.

If gold is predominant as $Au(HS)_2^-$ in ore fluids, gold solubility decreases with decreasing pH in reduced sulfur species (H_2S, HS^-) dominant region (Fig. 2.15). It seems likely that gold precipitates due to the mixing of gold-bearing hydrothermal solution and acid ground water and associated decreasing pH. Acid ground water occurs commonly in near-surface environment in geothermal system. This type of acid ground water is formed by sulfuric acids generated by the reaction

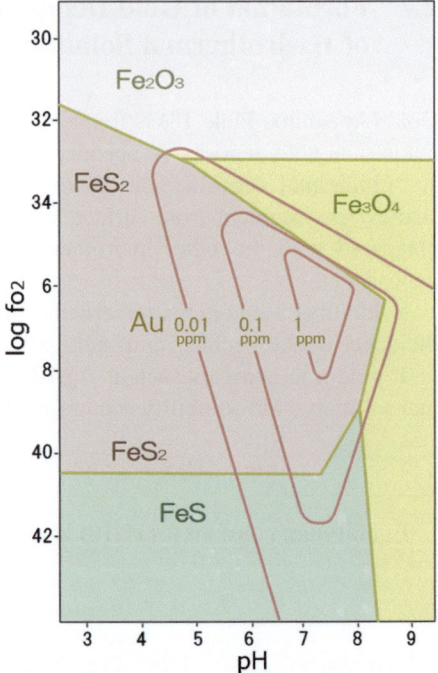

Fig. 2.15 $\log f_{O_2}$-pH diagram at 250 °C showing the stability fields of the principal sulfur species and solubility contours for gold in mg/kg · H_2O as $Au(HS)_2^-$

$$4SO_2(\text{volcanic gas}) + 4H_2O \rightarrow$$
$$3H_2SO_4 + H_2S \text{ and oxidation of } H_2S, H_2S + 2O_2 \rightarrow H_2SO_4 \qquad (2.21)$$

The amounts of minerals precipitated by the mixing of hydrothermal solution with acid ground water can be calculated, giving concentration and temperature of initial hydrothermal solution and acid ground water (Reed and Spycher 1984, 1985) (Fig. 2.16). The change in pH is important for the precipitation of gold that is obvious from the reaction

$$6H^+ + 8Au(HS)_2^- + 4H_2O \rightarrow 8Au(\text{gold}) + SO_4^{2-} + 15H_2S \qquad (2.22)$$

The mixing of hydrothermal solution with oxygenated ground water also causes the precipitation of gold (Shikazono 1993; Shikazono et al. 2002) (Fig. 2.17). As schematically shown in Fig. 2.18 gold tends to precipitate at near-surface environment due to the changes in pH, f_{O_2}, and temperature. Usually, pure gold does not precipitate from hydrothermal solutions. Electrum which is gold–silver alloy occurs in precious gold–silver deposits. Shikazono and Shimizu (1988) analyzed electrum from various types of ore deposits (epithermal Au–Ag vein-type, epithermal Pb–Zn vein-type (Appendix, Plate 24) Kuroko-type, skarn-type, mesothermal vein-type etc.) and estimated temperature, $\sum Au/\sum Ag$, and sulfur fugcity (f_{S_2}) and discussed the causes for electrum precipitation in these deposits.

2.7 Formation of Gold Deposits by Mixing of Hydrothermal Solution and Acid...

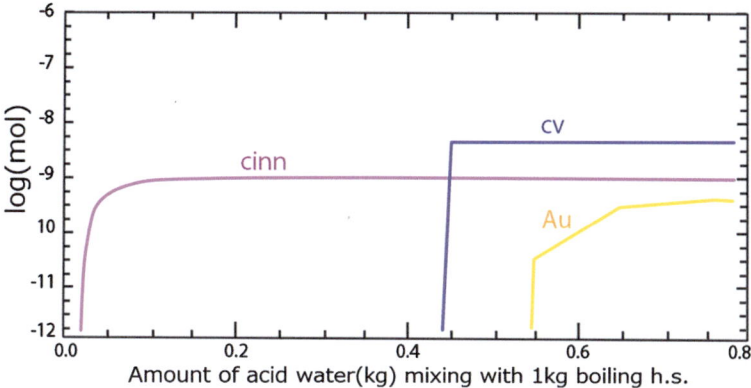

Fig. 2.16 Precipitated amounts of gold (Au), cinnabar (cinn, HgS), covellite (cv, CuS) by the mixing of acid sulfate solution and boiling hydrothermal solution (h.s.) (Reed and Spycher 1985)

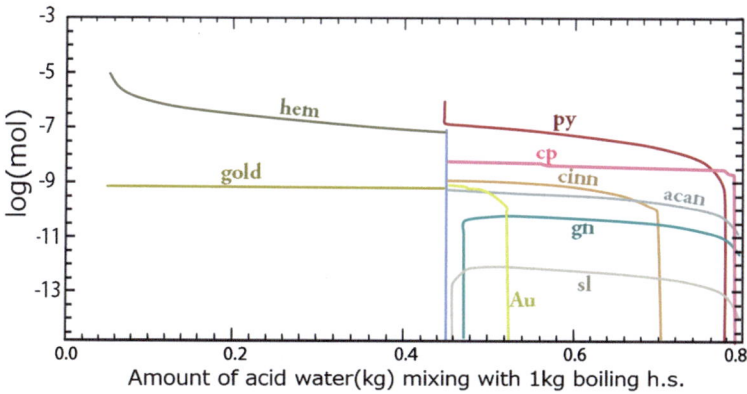

Fig. 2.17 Precipitated amounts of minerals by the mixing of boiling hydrothermal solution (h.s.) (100 °C) and ground water containing dissolved oxygen (Reed and Spycher 1985). *gold* native gold (Au), *acan* acanthite (Ag_2S), *cinn* cinnabar (HgS), *py* pyrite (FeS_2), *cp* chalcopyrite ($CuFeS_2$), *gn* galena (PbS), *sl* sphalerite (ZnS), *hem* hematite (Fe_2O_3)

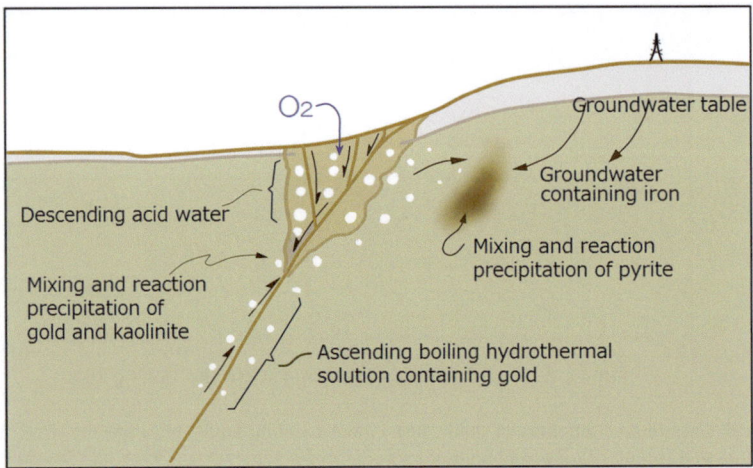

Fig. 2.18 Mixing and precipitation in upper part of hydrothermal system associated with boiling (Reed and Spycher 1985)

Cited Literature

Bischoff JL, Dickson FW (1975) Earth Planet Sci Lett 25:385–397
Drummond SE, Ohmoto H (1985) Geochim Cosmochim Acta 54:955–969
Giggenbach WF (1981) Geochim Cosmochim Acta 45:393–410
Helgeson HC (1979) Mass transfer among minerals and hydrothermal. In: Barnes HL (ed) Geochemistry of hydrothermal ore deposits. Wiley, New York, pp 568–610
Henley RW (1984) In: Reviews in economic geology, vol 1, pp 45–56
Holland HD, Malinin SD (1979) In: Barnes HL (ed) Geochemistry of hydrothermal ore deposits. Wiley, New York, pp 404–460
Mottl MJ (1983) Geol Soc Am Bull 94:161–180
Naumov GB, Ryzhenko BN, Khodakovsky IL (1974) Handbook of thermodynamic data. J S Geological Survey OSGS-WRD-74-001
Reed MH (1983) Econ Geol 78:446–485
Reed MH, Spycher NF (1984) Geochim Cosmochim Acta 46:513–528
Reed MH, Spycher NF (1985) In: Geology and geochemistry of epithermal system. Reviews in economic geol, vol 2, pp 249–272
Sakai H, Osaki S, Tsukagishi M (1970) Geochem J 4:27–39
Shikazono N (1974) J Fac Sci Univ Tokyo 19:27–56
Shikazono N, Holland HD (1983a) Econ Geol Mon 5:320–328
Shikazono N, Holland HD (1983b) Econ Geol Mon 5:329–344
Shikazono N (1984) Geochem J 18:181–187
Shikazono N (1985) Chem Geol 49:213–230
Shikazono N (1986) Min Geol 36:411–424
Shikazono N, Shimizu M (1987) Miner Deposita 22:309–314
Shikazono N (1988) Mining Geology Special Issue No. 12, pp 47–55
Shikazono N, Shimizu M (1988) Electrum: chemical composition, mode of occurrence and depositional environment. University of Museum, University Tokyo Bull No. 32
Shikazono N (1989) Chem Geol 76:239–247

Shikazono N, Kawahata H (1987) Can Min 25:465–474
Shikazono N, Shimizu M (1992) Can Min 30:137–143
Shikazono N (1993) Geochemistry 27:135–139 (in Japanese with English abstract)
Shikazono N, Nagayama T (1993) Resource Geology Special Issue No. 14, pp 47–56
Shikazono N (1994) Geochim Cosmochim Acta 58:2203–2213
Shikazono N, Nakata M, Tokuyama E (1994) Marine Geol 118:303–313
Shikazono N, Utada M, Shimizu M (1995) Appl Geochem 10:621–642
Shikazono N, Yonekawa N, Karakizawa T (2002) Res Geol 52:211–222
Shikazono N, Kawabe H, Ogawa Y (2012) Res Geol 62:352–368
Takeno N (1989) Min Geol 39:295–304
Taylor HP Jr (1997) In: Barnes HL (ed) Geochemistry of hydrothermal ore deposits, 3rd edn. Wiley, New York, pp 229–302
Thompson JB Jr (1959) In: Abelson PH (ed) Researches in geochemistry. Wiley, New York, pp 427–457
Truesdell AH (1984) In: Henley RW et al. (eds) Reviews in economic geology, vol 1, pp 129–142
Wolery TJ (1978) Some chemical aspects of hydrothermal processes at Mid-oceanic ridges- A theoretical study. I. Basalt-seawater reaction and chemical cycling between the oceanic crust and the oceans. II. Calculation of chemical equilibrium between aqueous solutions and minerals. Ph.D. thesis, Northwestern University, p 262
Wolery TJ (1983) User's guide and documentation – Livermore. Lawrence Livermore University, California, 191

Further Reading

Ellis AJ, Mahon WAJ (1977) Chemistry and geothermal systems. Academic, New York
Nishimura M (1991) Environmental chemistry Syokabo (in Japanese)
Ohmoto H, Skinner B J (eds) (1983) The Kuroko and related volcanogenic massive sulfide deposits. Economic Geology Monograph, vol 5
Prigogine I, Defay R (1954) Chemical thermodynamics (Everett DH trans). Jarrold and Sons, London
Shikazono N, Shimizu M (1988) Electrum: chemical composition, mode of occurrence and depositional environment. Wiley, New York
Shikazono N, Naito K, Izawa E (eds) (1993) High grade epithermal gold mineralization. Resource Geology Special Issue No. 15
Shikazono N (2003) Geochemical and tectonic evolution of arc-back arc hydrothermal systems: implication for the origin of kuroko and epithermal vein-type mineralization and the global geochemical cycle. Developments in geochemistry 8. Elsevier, Amsterdam
Walther JV, Wood BJ (1986) Fluid–rock interactions during metamorphism, vol 5, Advances in physical geochemistry. Springer, Berlin

Chapter 3
Mass Transfer Mechanism

Mass transfer during water–rock interaction process occurs sometimes under the condition close to equilibrium. However, generally at low temperatures it does not occur under near-equilibrium conditions.

Mass transfer mechanisms involve reaction, diffusion and advection. There are several types of reactions such as dissolution, precipitation, ion exchange (partitioning), adsorption and desorption.

Basic equation for the mass transfer involving above mechanisms is expressed as

$$\partial m_i / \partial t = \sum \Delta J\alpha - R_i \qquad (3.1)$$

where m_i is concentration of i species in aqueous solution, t is time, $\Delta J\alpha$ is divergence of flux such as advection, dispersion, and diffusion, and R_i is reaction-term.

At first, each mechanism (reaction, diffusion, and advection) will be explained and then coupling mechanisms will be applied to water–rock system (ground water system, hydrothermal system, seawater system).

3.1 Dissolution-Precipitation Kinetics

3.1.1 Dissolution Mechanism

Minerals dissolve into aqueous solution and precipitate from aqueous solution by the water–rock interaction process. The rate of dissolution of minerals for one component system is simply expressed as

$$dm_i/dt = k(m_{eq} - m_i)^n \qquad (3.2)$$

where m_i is concentration of i component in aqueous solution, m_{eq} is saturation (equilibrium) concentration of i component, k is rate constant (mass transfer coefficient), n is order of reaction, and $(m_{eq} - m_i)$ means a deviation from equilibrium concentration.

Rate constant (k) depends on mineral surface area/mass of water ratio, porosity, density of aqueous solution and temperature. Therefore, Eq. (3.2) converts into

$$dm_i/dt = k'(A/M)(1/\emptyset\rho)(m_{eq} - m_i)^n \quad (3.3)$$

where k' is rate constant of dissolution, A/M is mineral surface area(A)/mass of water (M) ratio, ø is porosity, and ρ is density of aqueous solution.

The other type of kinetic equation is given by

$$dm_i/dt = k''A/M\{1/\emptyset\rho\}]\{1 - (m_i/m_{eq})^m\}^n \quad (3.4)$$

where k'' is a rate constant of dissolution and m and n are constant values. Important variables governing the change in the concentration with time include k'', A/M, and m and n that will be explained below.

A/M: A/M is variable, depending on features of cracks (density, configuration, width etc.) and porosity. A/M of rock in which the crack with width of r occurs is r/2. A/M of rock consisting of closed packed grain sphere with diameter of r is $8.55V_{sp}/1,000r$ where V_{sp} is specific volume of water. BET method combined with porosity is conveniently used to estimate A/M. However, the results of these measurements are generally not in agreement with theoretical calculations by geometric method.

In general surfaces of mineral crystals do not homogeneously dissolve by the interaction of aqueous solution. The reactive surfaces at dislocation and imperfection sites tend to dissolve easily. The density of activated site significantly influences on dissolution and precipitation (Blum et al. 1990; Meike 1990).

It is likely that the rate of dissolution is simply proportional to A/M. However it has been reported that silicates in natural environment have higher A/M than that theoretically estimated, implying higher dissolution rate (Wells and Ghiorso 1991; Anbeek 1992).

k'', n: If rate-determining step is surface reaction, reaction rate varies significantly, depending on the kind of reaction product on the surface. The different reaction product is formed from aqueous solutions with different compositions. For instance, in the case of dissolution of quartz into pure water, the reaction, (Si–O–Si≡) + H_2O = (Si–O–Si·OH_2)*(*: transition state) → 2(≡Si–O–H), is a rate-determining step. In NaCl solution, chemical species shown in Fig. 3.1 are produced, which causes acceleration of reaction rates (Dove and Crerar 1990), indicating that the rate constant significantly depends on chemical composition of aqueous solution.

Above Eqs. (3.3 and 3.4) are established for diffusion-controlled rate determining step as well as for surface reaction-controlled one. In the case of

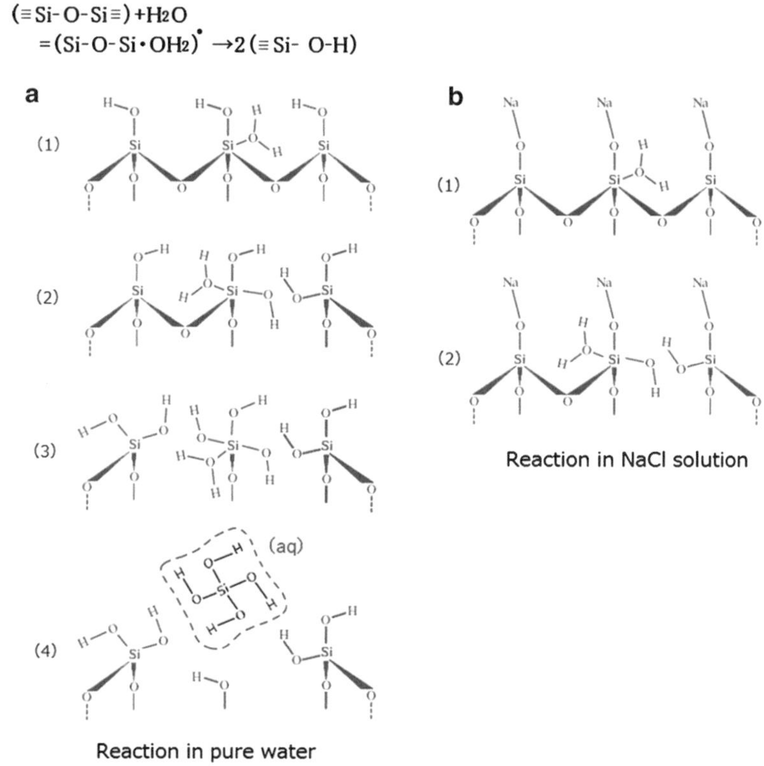

3.1 Dissolution–Precipitation Kinetics

$(\equiv Si\text{-}O\text{-}Si\equiv) + H_2O$
$= (Si\text{-}O\text{-}Si\cdot OH_2)^* \rightarrow 2(\equiv Si\text{-}O\text{-}H)$

Fig. 3.1 Dissolution process of quartz (Dove and Crerar 1990)

diffusion-controlled mechanism, k'' depends on diffusion coefficient and thickness of diffusion layer.

n for surface reaction-controlled mechanism (n = 2–5) is different from that for diffusion-controlled one (n = 1).

k'' varies with the style of advection. k'' for reaction-controlled mechanism does not change with the stirring rate, but that for diffusion controlled one does.

To determine which (diffusion or reaction) is the rate-determining step, there are the procedures such as (1) determination of n, (2) dependency of k' on stirring rate, (3) estimation of activation energy for the reaction (i.e., activation energy for diffusion-controlled-mechanism (-5 kcal mol^{-1}) is usually lower than that for surface reaction-controlled mechanism (-15 kcal mol^{-1})).

Rate-determining mechanism depends also on saturation index and pH. Rate-determining step for precipitation of salt minerals such as $BaSO_4$ from supersaturated aqueous solution is volume diffusion in aqueous solution, but surface reaction controls the precipitation when degree of supersaturation is small (Nielsen 1958).

Rate constants of dissolution and precipitation differ in different pH due to the involvement of H^+ and OH^- in the reactions (Fig. 3.2).

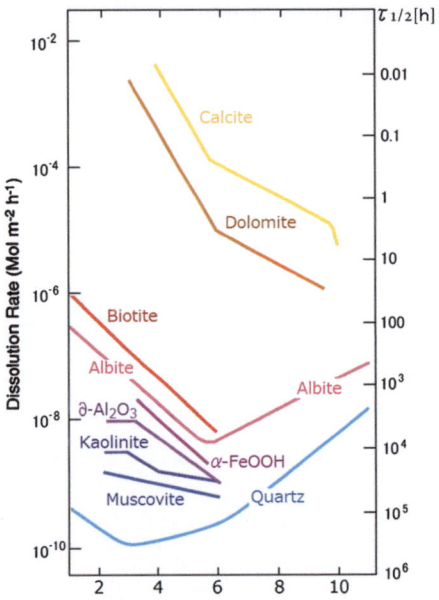

Fig. 3.2 Dissolution rate of different minerals (Sigg and Stumm 1993). On the right hand side halftime (hours) for the functional surface groups (S–OH, or S–CO$_3$H). Ten surface groups per nm^2 was assumed (Bidoglio and Stumm 1994)

The rate constants depend on temperature. Wood and Walther (1983) summarized the experimental results of dissolution rate of silicates as a function of temperature (Fig. 3.3). For the simple case of surface reaction the reaction rate is expressed as km (k: rate constant, m: concentration in aqueous solution), and for the diffusion-controlled mechanism, it is (D/x)m where D is diffusion coefficient and x is effective distance of diffusion. For the surface reaction mechanism, rate constant, k is Zexp (−E/kT) where E is activation energy and Z is constant value. Thus, reaction rate is Zexp (−E/kT)m.

If D = αT^2 (α: constant), for the diffusion-controlled mechanism, reaction rate = (αT^2/x)m. The temperature dependency of this reaction rate is shown in Fig. 3.3, where rate-limiting mechanism changes from reaction-controlled to diffusion-controlled with increasing temperature.

Rate equations, (3.2), (3.3) and (3.4), are not established for multi-component system. In this case, the rate of dissolution and precipitation of mineral is generally expressed as

$$dm/dt = k''(A/M)(1/\phi)\{(I.A.P./K_{eq})^m - 1\}^n = k''(A/M)(1/\phi)(\Omega^m - 1)^n \quad (3.5)$$

where $I.A.P.$ is ion activity product, K_{eq} is equilibrium constant of the dissolution reaction, Ω is degree of supersaturation that is $I.A.P./K_{eq}$, and m and n are constant for a given mineral.

The activation energy, m and n for precipitation reactions are given in Tables 3.1 and 3.2. It is obvious that reaction rate depends on solubility because I.A.P. relates to solubility.

3.1 Dissolution–Precipitation Kinetics

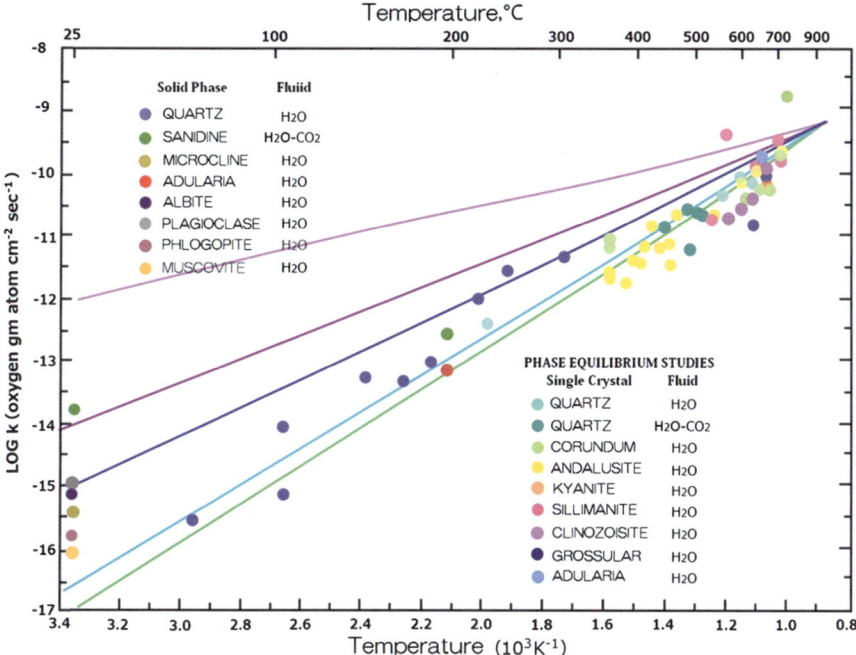

Fig. 3.3 Logarithm of the dissolution constant, k (oxygen gram atom cm^{-2} s^{-1}), plotted against the reciprocal of temperature (K^{-1}). The data are for near-neutral pH solutions. The *dashed line* gives the rate constant proposed by Wood and Walther (1983). The *solid lines* give the rate constants for the compensation law proposed in the text constructed for log k of -12, -14, -15, -17 at 25 °C (Wood and Walther 1983)

Table 3.1 Dissolution and precipitation rate parameters for mineral-solution reaction (Steefel and Cappellen 1990)

Mineral	Reaction	m	n
Kaolinite	$2Al^{3+} + 2SiO_2(aq) + 5H_2O \rightarrow Al_2Si_2O_5(OH)_4 + 6H^+$	1/9	2
K-feldspar	$KAlSi_3O_8 + 4H^+ \rightarrow K^+ + Al^{3+} + 3SiO_2(aq) + 2H_2O$	1	1
Gibbsite	$Al^{3+} + 3H_2O \rightarrow Al(OH)_3 + 3H^+$	1/4	2
Quartz	$SiO_2 \rightarrow SiO_2(aq)$	1	1
K-mica	$K^+ + 3Al^{3+} + 3SiO_2(aq) + 6H_2O \rightarrow KAl_2(AlSiO_3O_{10})(OH)_2 + 10H^+$	1/13	2
Halloysite	$2Al^{3+} + 2SiO_2(aq) + 7H_2O \rightarrow Al_2Si_2O_5(OH)_4 2H_2O + 6H^+$	1/9	2

Temperature and pH are not included as variables in Eq. (3.5). Rate equation as functions of temperature and pH is given by

$$Rate = k'''(A/M)\exp(-E/RT)a_{H^+}nH^+ \sum_i a_i^{n_i} f(\Delta G_r) \quad (3.6)$$

where k''' is rate constant, A/M is surface area/mass of aqueous solution ratio, E is activation energy, a is activity of i species, n is constant, and $f(\Delta G_r)$ is function of Gibbs free energy that deviates from equilibrium state.

Table 3.2 Activation energy for dissolution reaction (Lasaga 1984)

Mineral	Ea (kcal mol^{-1})	pH
Albite	13	Neutral
Albite	7.7	Alkaline
Albite	28	<3
Albite	17.1	1.4
Andalusite	11.5	1
Andalusite	5.8	2
Andalusite	1.8	3
Epidote	19.8	1.4
Kaolinite	16.0	1
Kaolinite	13.3	2
Kaolinite	10.3	3
Kaolinite	7.7	4
Kaolinite	2.3	6
Microcline	12.5	3
Prehnite	20.7	6.5
Prehnite	18.1	1.4
Quartz	17.0	7
Sanidine	12.9	3
Tephroite	12.9	2.5
Tephroite	6.3	3.5
Tephroite	5.7	4.2
Tephroite	1.1	5.1
Wollastonite	18.9	3–8

The dissolution rates of silicates such as gibbsite and kaolinite have been experimentally obtained (Mogollon et al. 1994; Ganor and Lasaga 1994; Nagy and Lasaga 1992). It is obvious in Eqs. (3.5) and (3.6) that the rate is related to degree of supersaturation. Concentrations and degree of supersaturation change with time in a closed system, indicating that precise rate constant and reaction rate are difficult to be estimated by a closed system experiment. In contrast, flow-through experiment in which concentration in a system is nearly constant is useful to determine rate constant and reaction rate (Lasaga 1998). Recently, the rate of dissolution has been investigated by microscopic examination of mineral surfaces, indicating the microscopic method is highly useful to determine the dissolution rate (e.g., Dove and Platt 1996; Sorai and Sasaki 2010).

3.1.2 Precipitation Mechanism

There are two important precipitation mechanisms including volume diffusion and surface reaction. Volume diffusion is diffusion of ions in aqueous solution towards the surface of minerals. Surface reaction means the precipitation reaction at crystal surface.

3.1 Dissolution–Precipitation Kinetics

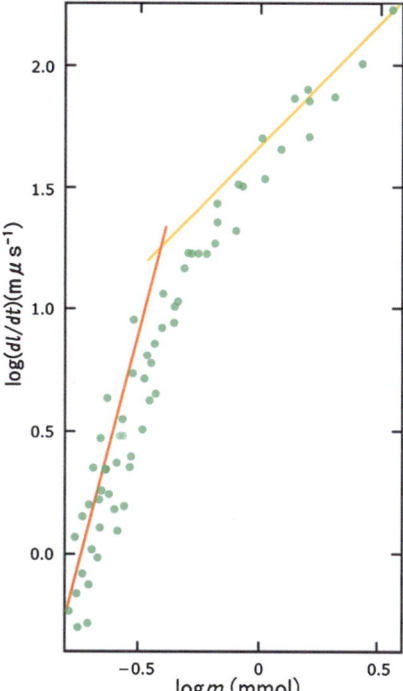

Fig. 3.4 Line growth rate as a function of $BaSO_4$ concentration (Nielsen 1958)

Surface reaction mechanisms include adsorption, desorption, surface nucleation, polynucleation, mononucleation and ion exchange reaction. The dependencies of amounts of precipitate and solution composition on time are different for each mechanism. For example, linear, exponential and logarithmic rate equations are established for volume diffusion, polynuclear growth and spiral growth, respectively.

In Table 3.1 n values previously obtained are given. n is 2–4 and 1 for low degree of supersaturation and for high degree of supersaturation, respectively. Nielsen (1958) experimentally showed n = 4 for the precipitation of $BaSO_4$ from the solution with low concentration (less than 0.4 m mol/L as $BaSO_4$) and n = 1 for high concentration (more than 0.4 mmol/L) (Fig. 3.4). Such kind of experimental studies indicated that volume diffusion and surface reaction (polynuclear growth) is dominant mechanism for the precipitation of salts from solutions with high concentration and low concentration, respectively.

Precipitation kinetics of salts such as barite ($BaSO_4$) has been studied to prevent the formation of scale ($BaSO_4$, $CaSO_4 \cdot 2H_2O$) in the pipes for the transportation of oil and geothermal water.

Nielsen and Toft (1984) discussed the precipitation mechanism using PA ($= -\log\sum A$) – PB($-\log\sum B$) diagram (Fig. 3.5). It is shown in this diagram that precipitation rate and mechanism depends on total cation and anion concentrations.

Calcite is the most well studied mineral with regard to precipitation kinetics (e.g., Inskeep and Bloom 1985; Shikazono and Shiraki 1994).

Fig. 3.5 PSO₄-PBa diagram for barite (BaSO₄) (Nielsen and Toft 1984)

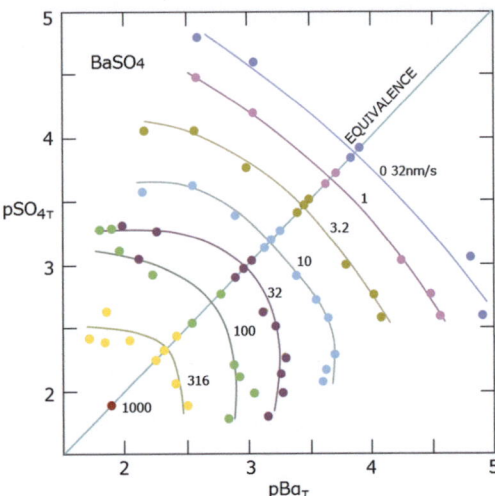

Davies and Jones (1955) proposed adsorption model. This is based on the following assumptions. (1) surface of crystal is enveloped by adsorption layer of hydrated ion, (2) crystallization of calcite occurs by dehydration reaction, (3) concentrations of cations and anions in adsorbed layer are constant if cation concentrations in solution are different. The precipitation rate equation is given by

$$R_{ppt} = -dm_{Ca^{2+}}/dt = k\left(m_{Ca^{2+}} - m_{Ca^{2+}aq}\right)\left(m_{CO_3^{2-}} - m_{CO_3^{2-}eq}\right) \quad (3.7)$$

where R_{ppt} is precipitation rate, m is concentration, k is precipitation rate constant, and m_{eq} is equilibrium concentration.

Electroneutrality relations are $m_{Ca^{2+}} = m_{CO_3^{2-}}$, $m_{Ca^{2+}eq} = m_{CO_3^{2-}eq}$. Therefore, we obtain

$$R_{ppt} = k\left(m_{Ca^{2+}} - m_{Ca^{2+}eq}\right)^2 \quad (3.8)$$

Reddy and Nancollas (1971) indicated (3.8) is established at 25 °C and pH > 8.8. If two ions concentrations are not low, Davies and Jones equation is given as follow. Two ions (A, B) concentrations in adsorbing layer are

$$\left.\begin{array}{l} m_{Asurface} = k'm_A exp(-\phi/RT) \\ m_{Bsurface} = k'm_B exp(\phi/RT) \end{array}\right\} \quad (3.9)$$

where ø is potential.

Above assumption and same concentration of two ions in adsorbing layer yield

$$exp(\phi/RT) = (m_A/m_B)^{1/2} \quad (3.10)$$

3.1 Dissolution–Precipitation Kinetics

Combining (3.10) with (3.8) we obtain

$$R_{ppt} = k\gamma^2 \left\{ (m_{Ca^{2+}})^{0.5} (m_{CO_3^{2-}})^{0.5} - (K_{sp}/\gamma^2) \right\}^2 \quad (3.11)$$

where K_{sp} is solubility product of calcite.

House (1981) and Kazmierczak et al. (1982) indicated this equation is established under pH > 7 and pH > 8.5. Putting degree of supersaturation given by $\Omega = a_{Ca^{2+}} a_{CO_3^{2-}}/K_{sp}$ into Eq. (3.11), we obtain

$$R_{ppt} = k\left(\Omega^{1/2} - 1\right)^2 \quad (3.12)$$

Reddy and Nancollas (1971) and Nancollas and Reddy (1971) showed that

$$R_{ppt} = k_p m_{Ca^{2+}} m_{CO_3^{2-}} \quad (3.13)$$

$$R_{dss} = k_d \quad (3.14)$$

where R_{dss} is dissolution rate, k_p and k_d are rate constants for precipitation and dissolution, respectively, R_{ppt} is precipitation rate, and R_{net} is $R_{ppt} - R_{dss}$. Therefore,

$$R_{net} = k_p \left\{ (m_{Ca^{2+}})(m_{CO_3^{2-}}) - K_{sp} \right\} \quad (3.15)$$

This equation corresponds to empirical rate equation, $R_{pp} = k(\Omega-1)^n$ for $n = 1$.

Plummer et al. (1979) conducted dissolution experiments for calcite at pH = 2–7, P_{CO_2} = 0.0003 – 0.97 atm and temperature = 50–60°C. They interpreted experimental results as follows; dissolution rate in pH < 3.5 does not depend on P_{CO_2}, in 3.5 < pH < 5.5 depends on both pH and P_{CO_2}, and in 5.5 pH does not on pH and P_{CO_2} but precipitation reaction is important. They considered that dissolution of calcite occurs by

$$CaCO_3 + H^+ = Ca^{2+} + HCO_3^- \quad (3.16)$$

$$CaCO_3 + H_2CO_3 = Ca^{2+} + HCO_3^- + H^+ + CO_2 \quad (3.17)$$

$$CaCO_3 + H_2O = Ca^{2+} + HCO_3^- + OH^- \quad (3.18)$$

They proposed the following model to explain experimental results.

$$R_{net} = k_1 a_{H^+} + k_2 a_{H_2CO_3} + k_3 a_{H_2O} - k_4 a_{Ca^{2+}} a_{HCO_3^-} \quad (3.19)$$

where k_1, k_2 and k_3 are rate constants for dissolution reaction (3.16), (3.17) and (3.18), respectively, and k_4 is precipitation rate constant. The theoretically derived k_4 is expressed as

$$k_4 = (K_{3\text{-}17}/K_{spc})(k_1 + 1/a_{H^+})(k_2 a_{H_2CO_3} + k_3 a_{H_2O}) \tag{3.20}$$

where $K_{3\text{-}17}$ is equilibrium constant for (3.17) and K_{spc} is solubility product for calcite. A few experimental studies on precipitation rate for calcite at high temperature have been conducted. Shiraki and Brantley (1995) reported experimental results on precipitation kinetics for calcite at 100 °C, 100 atm, and pH = 6.38–6.98. They indicated that in $a_{H_2CO_3} < 2.33 \times 10^{-3}$ and degree of supersaturation $(\Omega) < 1.72$ surface reaction controls precipitation rate of calcite and spiral growth on crystal surface is rate-limiting step, but in $\Omega > 1.72$ precipitation rate (R) is expressed as

$$R = A\exp\{-K/(\Delta G/RT)\} \tag{3.21}$$

and is controlled by two dimensional nucleation growth.

In $a_{H_2CO_3} < 2.33 \times 10^{-3}$ precipitation rate does not depend on $a_{H_2CO_3}$, but in $a_{H_2CO_3} > 2.33 \times 10^{-3}$ it significantly depends on $a_{H_2CO_3}$. In $a_{H_2CO_3} > 5.07 \times 10^{-3}$, precipitation rate is expressed as Eq. (3.15) and precipitation mechanism is adsorption of Ca^{2+} and CO_3^{2-} onto surface of calcite and in the condition where spiral growth is rate-limiting step, the model of Plummer et al. (1979) can be applied even in 100 °C.

3.1.3 Metastable Phase

It is frequently observed that precursor and metastable phase form before the formation of stable phase. For example, Schoonen and Barnes (1991) experimentally clarified that precursor, "FeS", is requisite for the formations of marcasite and pyrite (FeS_2). The changes from precursor through metastable phase to stable phase are commonly recognized. This process is called Ostwald ripening. Therefore, precipitation kinetics of precursor and metastable phase has to be elucidated.

Nielsen and Toft (1984) have experimentally obtained the stable field of precurser of calcite ($CaCO_3 \cdot 6H_2O$) on PA (PA = $-\log\Sigma A$) – PB (PB = $-\log\Sigma B$) diagram (Fig. 3.5). Degree of supersaturation for the solution from which the precursor forms is very high.

Although "FeS" formed by the experiments of Schoonen and Barnes (1991), its stability field is not determined.

Grain size of precursor and metastable phase is generally very small. Grain size (r*) relates to surface energy (σ) and degree of supersaturation (m/m_{eq}) as follows (Steefel and Cappellen 1990).

$$r^* = (2\sigma V_o/RT)\ln(m/m_{eq}) \tag{3.22}$$

where m is solubility of small grain, m_{eq} is solubility of large grain, V_o is molar volume of mineral, R is gas constant, and T is absolute temperature.

Surface energy of amorphous silica (46 mJ. m^{-2}) is considerably lower than that of quartz (335–385 mJ. m^{-2}). This difference in surface energy means that the grain size of amorphous silica is smaller than quartz which precipitated from same supersaturated solution. Critical concentration for the nucleation of amorphous silica is small. Thus, nucleation of amorphous silica occurs instead of quartz (Steefel and Cappellen 1990). Grain size of amorphous silica is very small. But it changes to quartz due to its crystal growth rate similar to quartz.

3.2 Diffusion

Diffusion involves (1) diffusion along grain boundary, and on surface of minerals, (2) diffusion of dissolved species in aqueous solution (volume diffusion, dispersion (convective diffusion), eddy diffusion, turbulent diffusion), and (3) diffusion in solid phase in the presence of aqueous solution.

3.2.1 Fick's Law

Basic equation (Fick's first law) of diffusion is expressed as

$$F = -Ddm/dz \,(g \cdot cm^{-2} \cdot s^{-1}) \quad \text{(Fick's first law)} \quad (3.23)$$

where F is flux, D is diffusion coefficient, m is concentration, and z is distance.

This is one dimensional flux in z direction (Lerman 1979).

Driving force for diffusion is a chemical potential difference. If flux does not change between z and z + Δz,

$$\Delta m/\Delta t = -\Delta F/\Delta z \,(g \cdot cm^{-3} \cdot s^{-1}) \quad \text{(t : time)} \quad (3.24)$$

Thus

$$(\partial m/\partial t)_z = -(\partial F/\partial z)_t \,(g \cdot cm^{-2} \cdot s^{-1}) \quad (3.25)$$

From (3.24) and (3.25), we obtain

$$\partial m/\partial t = \partial/\partial z (D \partial m/\partial z) \,(g \cdot cm^{-3} s^{-1}) \quad \text{(Fick's second law of diffusion)} \quad (3.26)$$

3.2.2 Diffusion in Pore in Rocks and Minerals

Diffusion in pore and crack in rocks and minerals is generally more important as micro scale mass transfer mechanism in water–rock system rather than diffusion in solid phases and in free water.

Diffusion coefficient of dissolved species in pore and crack is given by (Nakano 1991)

$$D_{cm} = \alpha\gamma D_o/\tau^2 \qquad (3.27)$$

where D_{cm} is molecular diffusion coefficient for solute in pore and crack, D_o is molecular diffusion coefficient in free water, α is correction coefficient for the variation of viscosity of water, γ is constant (less than 1) expressing the effect of interaction of ion in water and negatively charged surface of a grain, and τ is tortuosity which is the ratio of the mean length Lp of the path through the porous space between some two points to the straight line distance L between the same points ($\tau = Lp/L > 1$).

The following simplified equation is proposed.

$$D = D^*\alpha/\tau^2 \qquad (3.28)$$

where D^* is diffusion coefficient in infinite-dilute solution, α is viscosity of bulk water/viscosity of pore water ratio, $\tau^2 = \o F$, F is formation factor, $F = \o^{1-m}$, $F = R/R_0$, R (ohms) is electrical resistance of rock, and R_0 (ohms) is electrical resistance of pore water.

Apparent diffusion coefficient, D', when interactions such as adsorption occur, is given by

$$D' = (\o D_\o)/(\o + \rho Kd) = \o D_\o/(\o + \rho Kd)\delta/\tau^2 \qquad (3.29)$$

where ø is porosity, τ is tortuosity, ρ is density, Kd is distribution coefficient, and δ is constrictivity.

As mentioned above, diffusion coefficient depends on porosity, tortuosity and constriotivity of rocks and minerals. For example, effective diffusion coefficients of dissolved species in rocks and minerals correlate linearly with porosity.

3.3 Advection

Advection of aqueous solution in permeable rocks is an important mass transport mechanism as well as reaction and diffusion.

3.3.1 Darcy's Law

Advection of aqueous solution in permeable rocks can be generally treated as the solution which obeys Darcy's law. This is shown below.

3.3 Advection

Average velocity of ground water in porous media (rocks) is expressed as

$$q = -kadh/dx \quad (3.30)$$

where q is average volume flow rate, a is cross section area of permeable media, k is hydraulic conductivity (cms^{-1}), dh/dx is hydraulic gradient (h: height, x: distance) Generalized Darcy's law which is applicable to various liquid is expressed as (Tosaka 2006)

$$q = (Ka/\mu)s\delta\Psi/\delta x \quad (3.31)$$

$$v = q/a = -(K/\mu)\delta\Psi/\delta x \quad (3.32)$$

where v is average velocity, μ is viscosity coefficient of fluid, K (m^2) is permeability, Ψ is fluid potential. The fluid potential is given by

$$\Psi = P + \rho g h \quad (3.33)$$

where Ψ is fluid potential, P (Pa) is pressure, ρ is density of fluid, h is height and g is the gravitational acceleration.

The relation between hydraulic conductivity (κ) and permeability (K) is

$$\kappa = \rho g K/\mu \quad (3.34)$$

If the pressure difference between 1 cm in porous media is 1 atm, viscosity of fluid is 1 cp, and volume flow rate is 1 cc. s^{-1} for the cross section with 1 cm^2, is 1 darcy. This Darcy's law is applicable to laminar flow with Reynolds number (Re) less than 1–10.

Reynolds number, Re, is expressed as

$$Re = 2vr/\nu \quad (3.35)$$

where v is velocity of fluid, r is grain size, and ν is kinematic viscosity coefficient (ρ/μ).

The flow becomes to be turbulent under Re more than 1–10, implying that Darcy's law is not established.

Above equations indicate that flow rate is determined by permeability, viscosity and hydraulic gradient. Permeability of rocks varies very widely (Fig. 3.6). It ranges from 10^{-15} to 10^{-3} cm^2 (10^5 Darcy). It varies widely even in a same kind of rock. For example, permeability of sediments is in a range of 10^{-15} cm^2 (10^{-7} Darcy) to 10^{-3} cm^2 (10^{-5} Darcy) (Fig. 3.6).

Permeability is estimated experimentally and theoretically (Lerman 1979). It is related to sorting, porosity, pore size, specific surface area, configuration of grain and grain size. Particularly, the relationship between permeability and porosity and width of fracture has been studied in detail. For example, the following relationship between permeability and width of fracture is proposed (Phillips 1991).

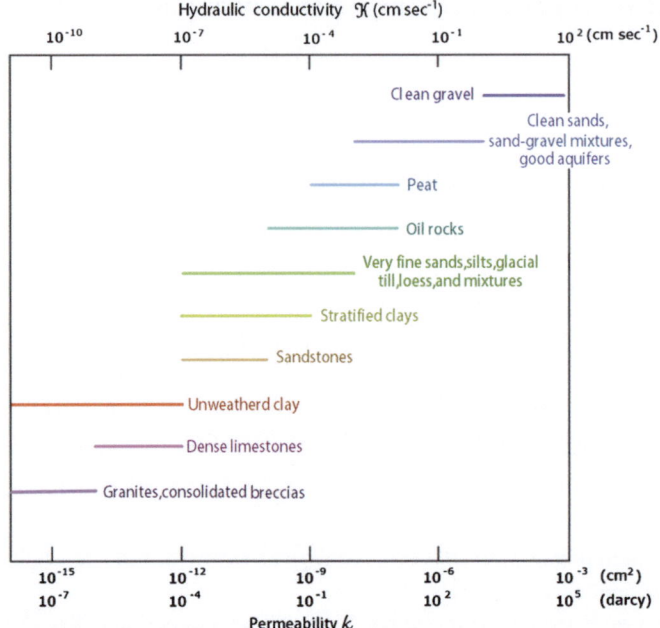

Fig. 3.6 The coefficients of hydraulic conductivity and permeability of sediments and rocks (from numerous literature sources)

$$K = \emptyset \delta^3 / 12 \tag{3.36}$$

where K is permeability, $\emptyset$ is porosity, and δ is width of fracture.

Permeability depends on the types of pore. Various types of pore exist in natural rocks. They are divided into primary and secondary pore. Primary pore includes pore in grains, pore between grains and vesicle. Secondary pore includes pore formed by leaching and fracturing (Uchida 1992). Permeability changes by compaction and formation of secondary minerals at diagenetic stage. However, the change in permeability with time and dependency of permeability on types of pore have not been investigated.

3.3.2 Three Dimensional Fluid Flow

Darcy's law for three dimensional fluid flow is given by

$$v_x = k(x, y, z) \partial P / \partial x \tag{3.37}$$

where v_x is velocity in x direction, $k(x, y, z)$ is hydraulic conductivity.

Continuity equation for steady state fluid flow is

$$\partial v_x/\partial x + \partial v_y/\partial y + \partial v_z/\partial z = 0 \qquad (3.38)$$

Combination of Eq. (3.37) with (3.38) leads to

$$\partial/\partial x[k(x,y,z)\partial P/\partial Dx] + \partial/\partial y[k(x,y,z)\partial P/\partial y] + \partial/\partial z[k(x,y,z)\partial P/\partial z] = 0 \qquad (3.39)$$

If k is constant, (3.39) is converted into

$$\partial^2 P/\partial x^2 + \partial^2 P/\partial y^2 + \partial^2 P/\partial z^2 = 0 \qquad (3.40)$$

$$\nabla^2 P = 0 \qquad (3.41)$$

Giving boundary conditions and solving this equation, three dimensional iso-potential contours in underground can be obtained. From the contours, we know three dimensional ground water flow whose direction is perpendicular to iso-potential contours.

Above treatment is simplified one. However, natural system is more complicated. For instance, fluids are compressive, fluids flow under non-steady state conditions, and permeability of rocks is heterogeneously distributed. Fluids flow not only in saturated zone, but also in unsaturated zone.

3.4 Coupled Models

Above-mentioned reaction, diffusion and advection influence mass transfer in rock-water system. It is generally difficult to solve the differential equation including all these mechanisms. Thus, the two coupled models at constant temperature and pressure will be explained below. They are (1) reaction-fluid flow model, (2) reaction-diffusion model, (3) diffusion-fluid flow model. In addition to these coupled models, model taking into account the change in temperature will be considered.

3.4.1 Reaction-Fluid Flow Model

Modeling of mass transfer has been carried out in the field of chemical engineering and environmental engineering (e.g., Takamatsu et al. 1977). Models commonly used are (1) batch model, (2) perfectly mixing model (Fig. 3.7), (3) piston flow model (Fig. 3.8), and (4) tank model (multi-step model).

Batch model (closed system model) is described in Sect. 3.1. (2) and (3) will be described below.

Fig. 3.7 Perfectly mixing model

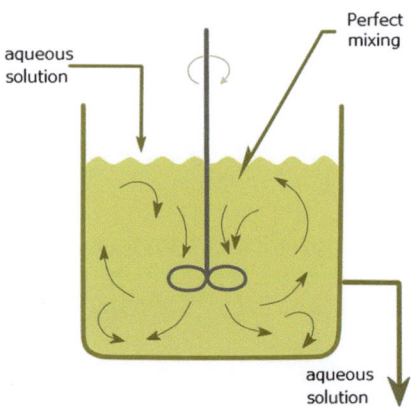

Fig. 3.8 Piston flow model (Fujimoto 1987)

3.4.1.1 Perfectly Mixing Fluid Flow-Reaction Model

Perfectly mixing fluid flow-reaction (Fig. 3.7) for one component system is described as

$$dm/dt = k(m_{eq} - m)^n + (q/V)(m_i - m) \quad (3.42)$$

where m is concentration, k is reaction rate constant, m_{eq} is saturation (equilibrium) concentration, q is volume flow rate, m is initial concentration, V is volume of aqueous solution in system and n is empirical and dimensionless parameter.

Assuming that n is equal to 1, and $dm/dt = 0$, Eq. (3.42) becomes to

$$k(m - m_{eq}) + (q/V)(m_i - m) = 0 \quad (3.43)$$

Therefore, we obtain

$$m = \{m_{eq} + qm_i/(kV)\}\{1 + q/(kV)\} \quad (3.44)$$

This indicates concentration, m is governed by m_{eq}, m_i and $q/(kV)$.

This simple model is applied to chemical weathering of silicates caused by the interaction of rainwater penetrating permeable rocks downward to interpret Goldich's weathering series.

3.4 Coupled Models

We consider the simple dissolution reaction from solid A (A_s) to A in aqueous solution (A_{aq}). Concentration of A in aqueous solution is governed by dissolution of solid A and outflow of A in aqueous solution from the system. Dissolution rate of solid is given by

$$dA_s/dt = k(A_{eq} - A) - (q/V)A \tag{3.45}$$

In a case of steady state

$$A = A_{eq}/\{1 + q/(kV)\} \tag{3.46}$$

If dominant cation and anion is B and HCO_3^-, respectively, it can be approximated as

$$B_{eq} = m_{HCO^-} \tag{3.47}$$

where B_{eq} is equilibrium (saturation) concentration of B.

Exchange reaction is written as

$$BR + A_{eq} = AR + B_{eq} \tag{3.48}$$

where BR and AR are solid phases.

Equilibrium constant for ion exchange reaction of (3.47) is expressed as

$$K_{3\text{-}47} = (B_{eq}/A_{eq}) \tag{3.49}$$

where a_{BR} and a_{AR} are assumed to be unity.

Combination of (3.49) with (3.46) leads to

$$A_{eq} = m_{HCO_3^-}/K_{3\text{-}47} \tag{3.50}$$

This equation and (3.46) yield

$$A = \{m_{HCO_3^-}/K_{3\text{-}47}\}/\{1 + q/(kV)\} \tag{3.51}$$

Equation (3.51) implies that A is determined by $q/(kV)$ and equilibrium concentration. Deviation from the equilibrium depends on $q/(kV)$. Equilibrium concentration is related to $a_{H_2CO_3}$ (P_{CO_2}) and pH. If $q/(kV)$ is considerably lower than 1, (3.51) is expressed as

$$A = A_{eq} \tag{3.52}$$

This equation represents equilibrium condition. If $q/(kV)$ is considerably larger than 1, we obtain

Table 3.3 Dissolution rate constants of silicates

Mineral	Dissolution rate of Si (mol m^{-2} s^{-1})
Quartz	4.1×10^{-14}
Muscovite	2.56×10^{-13}
Forsterite	1.2×10^{-12}
K-feldspar	1.67×10^{-12}
Na-feldspar	1.19×10^{-11}
Enstatite	1.0×10^{-10}
Diopside	1.4×10^{-10}
Nepheline	2.8×10^{-10}
Ca-feldspar	5.6×10^{-9}

$$A = A_{eq}k(V/q) = A_{eq}k\tau \quad (3.53)$$

where τ ($=V/q$) is residence time.

τ is not different for different mineral. Thus, (3.51), (3.52) and (3.53) indicate that chemical weathering series of silicates depend on dissolution rate constant and equilibrium concentration. The dissolution rate constants of silicates are summarized in Table 3.3 which indicates that the rate constant and solubility (equilibrium concentration) increase as the following order; quartz → K-feldspar → Na-feldspar → enstatite → diopside → Ca-feldspar. This order is consistent with Goldich's weathering series (Goldich 1938).

3.4.1.2 Piston Flow-Reaction Model

Piston flow-reaction model (Fig. 3.8) for one component and one dimensional system is expressed as (Fujimoto 1987; Shikazono and Fujimoto 1996)

$$\partial m/\partial t + v\partial m/\partial x = (Ak/M)(m_{eq} - m)/m_{eq} \quad (3.54)$$

where m is concentration, v is velocity, x is distance, A/M is surface area of rock(A) / mass of aqueous solution (M) ratio, k is reaction rate constant, and m_{eq} is equilibrium concentration.

If $\partial m/\partial t = 0$ (steady state condition), concentration is expressed as

$$m = m_{eq} - (m_{eq} - m_i)exp\{-Akx/Mvm_{eq}\} \quad (3.55)$$

where m_i is input initial concentration.

If m_i is negligible compared with m_{eq}, (3.55) is given by

$$m/m_{eq} = 1 - exp\{-Akx/Mvm_{eq}\} \quad (3.56)$$

Chemical composition of ground water in granitic rock area (Tsukuba, Ibaraki Prefecture, Japan) is interpreted based on Eq. (3.56) (Shikazono and Fujimoto

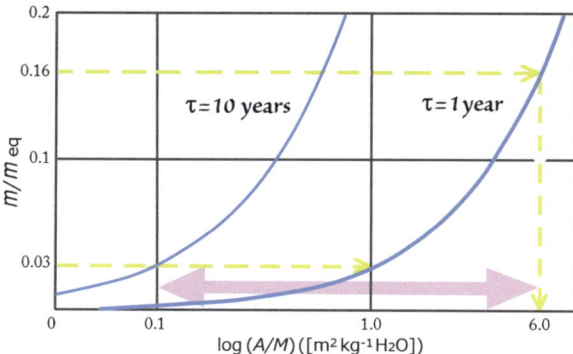

Fig. 3.9 Relationship between m/m$_{eq}$ and A/M based on perfectly mixing model (Shikazono and Fujimoto 1996). m concentration, m_{eq} equilibrium concentration, A surface area of mineral, M mass of aqueous solution, τ residence time

2001). SiO$_2$ concentration of ground water in Tsukuba granitic rock area, central Japan is in a range of $(1.7–5.0) \times 10^{-4}$ mol · kg H$_2$O^{-1} which is lower than equilibrium concentration.

Shikazono and Fujimoto (2001) used a coupled fluid flow dissolution kinetics model (Eq. (3.54)) to interpret the SiO$_2$ concentration of ground water in Tsukuba granitic rock area (Japan). Assuming reasonable values of residence time, equilibrium concentration of SiO$_2$ in ground water, and dissolution rate of Na·Ca-feldspar, A/M was estimated to be 0.15–8.7 (Fig. 3.9). The width of a fracture in a granitic rock is estimated to be 0.2 mm^{-2} cm. The estimated value of the width is much wider than that obtained by pore sizes experimentally determined using the mercury intrusion method for a hand-specimen scale granitic samples, which is $n \times 10^{-1}$ μm. This suggests that the interaction of ground water and country rocks takes place dominantly on the surface of a fracture with a width of ca. 0.2 mm^{-2} cm and the chemical composition of ground water is controlled by such interaction and fluid flow in the fracture.

Silica concentration in deep ground water in the granitic rock area (e.g., Kamaishi, Japan) is in equilibrium with SiO$_2$ mineral (chalcedony) (Fig. 1.27). Based on a coupled fluid flow-dissolution-precipitation kinetics model the relationship between residence time of deep ground water and A/M was derived, and the reasonable values of τ is estimated to be more than 40 years (Shikazono and Fujimoto 2001).

Above calculation indicates that dissolution rate constant experimentally determined can explain the SiO$_2$ concentration in ground water. However, previous studies showed that the estimate of dissolution rate constant from field studies is not in agreement with experimental dissolution rate constant (Table 3.4). It is suggested that the disagreement is due to that the estimate of A/M by field studies is not correct. However, there have been considerable discussions on this discrepancy between field and laboratory dissolution rate constants. This difference is considered to be caused by the influences of solute composition (e.g., Al^{3+}), thermodynamic saturation states, A/M ratio, solution/solid ratios, secondary surface coating, compositional differences within single mineral grains and others (White 2004).

Table 3.4 Comparison of laboratory and field weathering rates

Mineral	Laboratory weathering rate (mol Si m^{-2} s^{-1})	Field-estimate weathering rate (mol Si m^{-2} s^{-1})	Field-measured cation export (equiv ha^{-1} year^{-1})	Cation(s)	Notes
Plagioclase (oligoclase)	5 × 10^{-12}	3 × 10^{-14}	210	Na$^+$	Trnavka River Basin (CZ)
Plagioclase (oligoclase)	5 × 10^{-12}	8.9 × 10^{-13}	350	Na$^+$, Ca^{2+}	Coweeta Waterdhed, NC (USA)
Almandine Biotite	5 × 10^{-12}	3.8 × 10^{-12}	300	Mg^{2+}, Ca^{2+} K$^+$, Mg^{2+}	Filson Creek, MN (USA)
Plagioclase (bytownite)		1.2 × 10^{-13}	150	Ca^{2+}, Na^{2+}	
		5 × 10^{-15}	330	Mg^{2+}	
Olivine	7 × 10^{-12}	1 × 10^{-13}	310		
Plagioclase epidote, biotite		6 × 10^{-14}	200	Ca^{2+}, Na$^+$, K$^+$	Cristallina, Switzerland
Plagioclase, biotite	6 × 10^{-12}	9 × 10^{-15}	960	Ca^{2+}, Na$^+$, Mg^{2+}	Bear Brooks Watershed, Maine (USA) Assumption: 50 cm of saturated regolith, 0.5 m^2 g^{-1} surface area of mineral grains measured, 40 % of mineral grains active in weathering

3.4.1.3 Non-steady State Perfectly Mixing Fluid Flow-Reaction Model

Dissolution of mineral is considered based on non-steady state perfectly mixing-fluid flow-reaction model by Fujii (1977). The change in concentration in aqueous solution by the dissolution of solid phase is given by

$$dm/dt = f(m_i - m) + k(m_{eq} - m) \tag{3.57}$$

where $f = q/V = 1/\tau =$ constant, q is volume flow rate, V is volume of water in a system, and τ is residence time.

m_i/m_{eq} and m are substituted by a and y, respectively (Kitahara 1960; Wollast 1974). This gives

$$dy/dt = k\{(f/k)a + 1\} - \{y[(f/k) + 1]\} \tag{3.58}$$

Thus, we obtain

$$y = \{(f/k)a + 1\}/\{(f/k) + 1\} \\ - [\{(f/k)a + 1\}/\{(f/k) + 1\} - v_0]exp\{-(k+f)t\} \tag{3.59}$$

If $t = \infty$ and $y = y_\infty$, y_∞ becomes

$$y_\infty = \{(f/k)a + 1\}/\{(f/k) + 1\} \tag{3.60}$$

Dissolution rate of solid phase is written as

$$-dz/dt = f\{a - (z_0 + v_0) + z\} + k\{1 - (z_0 + v_0) + z\} \tag{3.61}$$

where $z = x/X$, X is mass of solid dissolved in the saturated solution, x is mass of solid after time, t. z_0 is $z_0 = x_0/x$, where x_0 is initial mass of solid, and $v_0 = u_0/x$, where u_0 is initial mass of solid dissolved.

If $z = (z_\infty)_{flow}$, where $(z_\infty)_{flow}$ is z for $t=\infty$

$$(z_o)_{flow} = (z_o + v_o) - \{(f/k)/a + 1\}/\{(f/k) + 1\} \tag{3.62}$$

In a closed system, $(z_o)_{closed} = z_o + v_o - 1$. Thus

$$(z_0)_{flow} - (z_\infty)_{closed} = (f/k)(1 - a)/\{(f/k) + 1\} \tag{3.63}$$

This equation indicates that the concentration in flow system is controlled by f/k and a.

According to Margenau and Murphy (1956)

$$-dz/dt = f\{a - (z_0 + v_0) + z\} + kz(w_0 + z) \tag{3.64}$$

Fig. 3.10 Decrease of concentration as a function of f/k (Fujii 1977). *LE* logistic equation model, *KW* Kitahara and Wollast model, *MM* Margenau and Murphy model (Margenau and Murphy 1956)

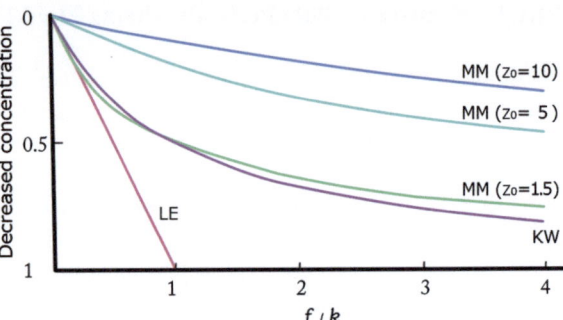

If $\{(f/k) + w_0\}^2 > 4(f/k)(a - z_0 - v_0)$, the solution is

$$C' \exp\left[\left\{-k((f/k) + w_0)^2 - 4(f/k)(a - z_0 - v_0)\right\}^{1/2} t\right] \quad (3.65)$$

where C' is constant that is expressed as

$$C' = \left[2z_0 + \{(f/k) + w_0\} - \{(f/k) + w_0\}^2 - 4\{(f/k)(a - z_0 - v_0)\}^{1/2}\right] \Big/ \left[\{2z_0 + ((f/k) + w_0)^2 + ((f/k) + w_0)\}^2 - 4(f/k)(a - z_0 - v_0)\right]^{1/2} \quad (3.66)$$

If $t \to \infty$, $v_0 = 0$ and $a = 0$

$$(z_\infty)_{flow} - (z_\infty)_{closed}$$
$$= (1/2)\{(f/k) + 1 - z_0\}^2 + \{4(f/k)z_0\}^{1/2} - (1/2)\{(f/k) + z_0 - 1\}\} \quad (3.67)$$

According to logistic model*

$$-dz/dt = f[a - (z_0 + v_0) + z] + k[(z_0 + v_0) - z][1 - (z_0 + v_0) + z] \quad (3.68)$$

If $a = 0$ and $v_0 = 0$, this yields

$$(z_\infty)_{flow} - (z_\infty)_{closed}$$
$$= (1/2)\{(f/k) + 1\} - \left[\{1 - 2z_0 - (f/k)\}^2 - 4z_0((f/k) + z_0 - 1)\right]^{1/2} \quad (3.69)$$

The different relations shown in Fig. 3.10 indicate that the three models are dependent on the different dissolution mechanism and kinds of reactions involved.

3.4 Coupled Models

*Logistic model:

Logistic model is expressed as

$$dN/dt = N(a - bN) \tag{3.70}$$

where N is number, a and b are constant. This is solved as

$$N = (a/b)/\{1 + (a - bN_0)/(bN_0)\}\exp(-t) \tag{3.71}$$

3.4.1.4 Non-steady State Piston Flow-Reaction Model

Non-steady state piston flow-reaction model for multi-component system was applied to alteration and weathering of rocks by Lasaga (1984). Lasaga (1984) calculated the changes in water and rock compositions during the weathering of feldspar and nepheline ($Na_3(Na,K)(Al_4Si_4O_{16})K(AlSiO_4)$) to gibbsite ($Al(OH)_3$) and kaolinite ($Al_2Si_2O_5(OH)_4$) based on fluid-flow reaction model which is expressed as

$$\partial(\phi m_i)/\partial t = \phi R_i - \partial(\phi v_x m_i)/\partial x \tag{3.72}$$

where ϕ is porosity, R_i is rate of dissolution and precipitation, x is distance and v_x is flow rate (in x-direction).

R_i is obtained from

$$dm_i/dt = (A_Q/V)v_{\theta i}k_+(a_{H^+})^{n\theta} - (A_\theta/V)v_{\theta i}(Q^m/K^{rn}_{eq})k_+(a_{H^+})^{n\theta} \tag{3.73}$$

where A_Q/V is surface area of mineral/volume of aqueous solution ratio, Q is the reaction activity quotient.

The vertical profile of the Si and Al concentrations from surface was calculated using (3.72) and (3.73) and is shown in Fig. 3.11.

3.4.2 Reaction-Diffusion Model

If two minerals contact with each other at constant the pressure-temperature condition where two minerals are unstable, reaction occurs between them to form stable mineral. The dominant rate limiting mechanisms are diffusion of aqueous species dissolved from minerals in fluid and dissolution and precipitation reactions. If fluid is not present, diffusion in solid phase occurs. But the rate of diffusion in solid phase is generally very slow. However, at very high temperature and pressure (metamorphic condition) the diffusion in solid phase may control the mass transfer. Reaction-diffusion model is able to be used to obtain the development of reaction zone between two minerals with time.

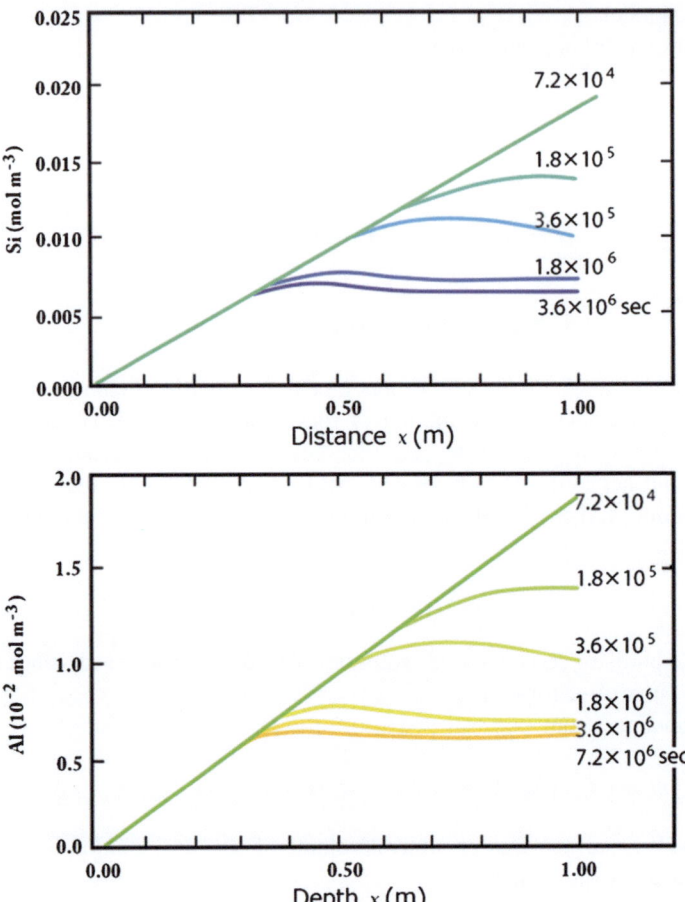

Fig. 3.11 The change in Si and Al concentrations in ground water with depth from surface (Lasaga 1984)

For example, formations of talc ($Mg_6Si_8O_{20}(OH)_4$) and serpentine ($Mg_3Si_2O_5(OH)_4$) between quartz and forsterite(Mg-olivine) (Mg_2SiO_4) which initially contact with each other are calculated based on reaction-flow model (Fig. 3.12) (Lichtner et al. 1996). Abundances of talc and serpentine increase but those of quartz and forsterite decrease with time. Between forsterite and serpentine, tremolite and chlorite form (Fig. 3.13) (Nishiyama 1987).

3.4.3 Diffusion-Flow Model

Hydrothermal solution issues from mid-oceanic ridges and back arc basins. Sulfates and sulfides precipitate from hydrothermal solution, resulting to the formation of chimney (Fig. 3.14). Mineral assemblages vary from outer to inner parts of the

3.4 Coupled Models

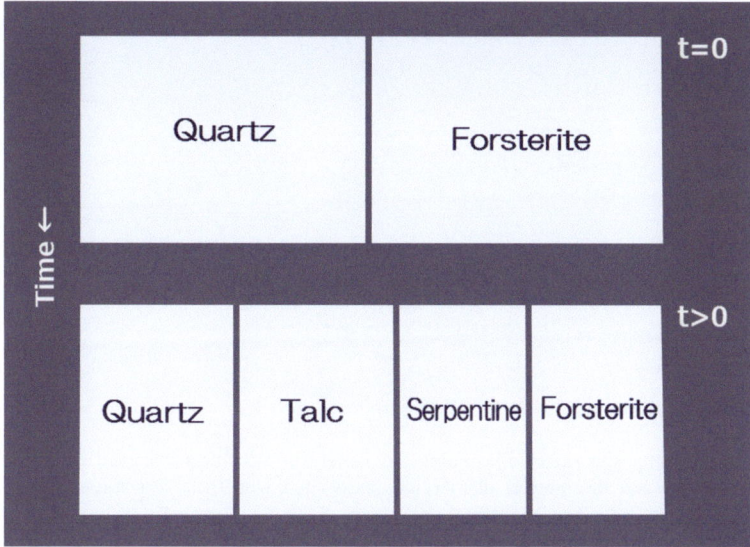

Fig. 3.12 Formation of talc and serpentinite between forsterite and quartz which contract initially (Lichtner et al. 1996)

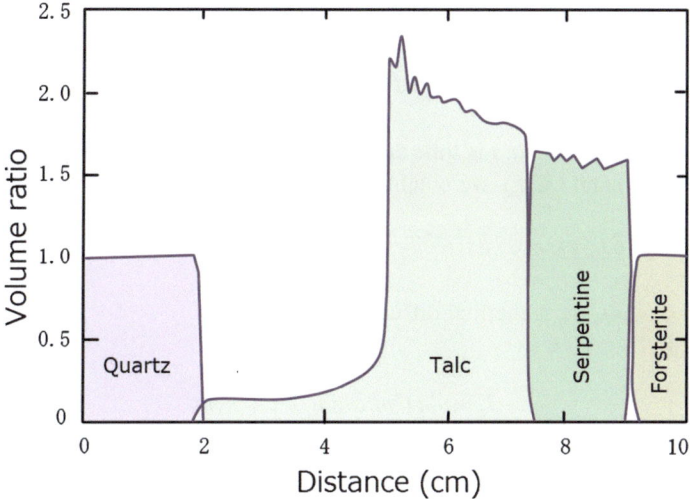

Fig. 3.13 Volume ratio of minerals away from quartz which form during $(2–5) \times 10^5$ s in $MgCl_2$–SiO_2–H_2O–HCl system (390 °C) (Lichtner et al. 1996)

chimney that is Cu-, Fe-sulfides $\rightarrow$ Zn-, Fe-sulfides $\rightarrow$ sulfates. Tivey and McDuff (1990) interpreted this zoning using diffusion-flow model as follows.

Heat flux through chimney is given by

$$J_x = -\kappa \partial T/\partial x + (\rho C_p)_l \varnothing v T \tag{3.74}$$

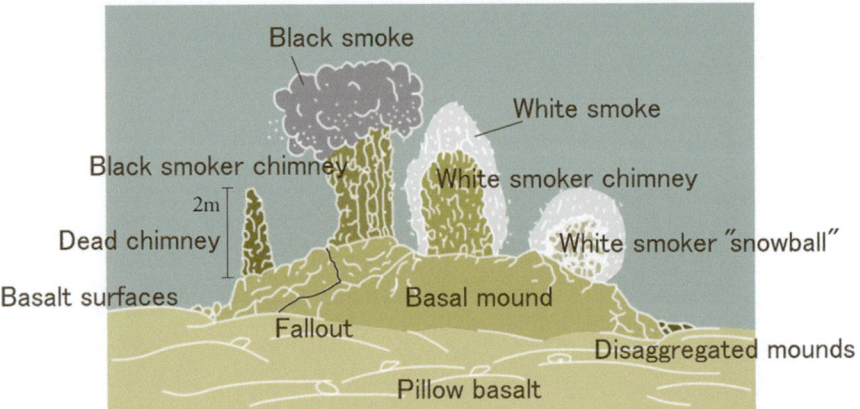

Fig. 3.14 A composite sketch illustrating the variety of structures observed at the different RISE vent sites and the mineral distributions associated with these structures (Haymon and Kastner 1981)

where J_x is heat flux in x direction, κ is heat conductivity, T is absolute temperature, x is distance, ρ is density of fluid, C_p is heat capacity, l is fluid, ø is porosity of chimney, and v is average velocity of fluid.

Temperature is expressed as a function of x as

$$(\rho C_p)_m \partial T/\partial t + \partial J_x/\partial x = 0 \quad (3.75)$$

where m is porous media, t is time and J_x is heat flux of x direction.

From (3.74) and (3.75), we obtain

$$\partial T/\partial t = \partial/\partial x(K\partial T/\partial x) - \text{ø}u(\rho C_p)_l/(\rho C_p)_m \partial T/\partial x \quad (3.76)$$

where $K = \kappa/\rho(C_p)_m$ is thermal diffusibility.

This can be solved as

$$J_i = -D_i' m_i \partial \ln a_i/\partial x + \text{ø}v m_i \quad (3.77)$$

where J_i is flux of dissolved i species, D_i' is diffusion coefficient for sediments, a is activity, and m is concentration.

This equation is converted into

$$J_i = -D_i \partial m_i/\partial x_i - D_i' m_i \partial \ln \gamma_i/\partial x + \text{ø}v m_i \quad (3.78)$$

where γ_i is activity coefficient.

Hydrothermal solution ascends in chimney and migrates through the chimney. Temperature distribution in chimney is determined by the migration of hydrothermal solution. The temperature distribution in chimney is obtained from the following equations.

$$(\rho C_p)_m \partial T/\partial t + \partial J_x/\partial x = 0 \qquad (3.79)$$

$$J_x = -\kappa \partial T/\partial x + (\rho C_p)_l \emptyset vT \qquad (3.80)$$

It is also possible to estimate the spatial distribution of steady state concentration in chimney, giving temperature dependency of diffusion coefficient, D. The amounts of precipitation of minerals can be calculated from the saturation index that is obtained from the distribution of concentration of fluid in chimney.

3.4.4 Temperature-Dependent Model

Temperature gradient in geothermal and hydrothermal systems is generally very steep. Thus, it is necessary to consider the effect of temperature gradient on mass transfer. Temperature affects significantly reaction rate constant, diffusion coefficient, and fluid flow.

Wells and Ghiorso (1991) calculated the variation in SiO_2 concentration in hydrothermal solution ascending from deeper part, accompanied by a decrease in temperature in mid-oceanic ridge hydrothermal system based on the following one component steady state formation.

$$\partial m_i/\partial t = \sum_{j=1}^{M} (A/V)_{v_{ij}} k(a_{H^+})^n \{1 + \exp(n\Delta G_0/RT)\} - \partial V_x/\partial x \qquad (3.81)$$

where m_i is concentration of i component, A is surface area of mineral contacting with aqueous solution, k is reaction rate constant, V is volume of water reacting with mineral, v_{ij} is stoichiometric coefficient of i component in mineral j, ΔG_j is Gibbs free energy of reaction, n is constant, v_x is velocity, D_i is diffusion coefficient of i component, a_i is activity of i component, and x is distance.

They indicated that there can be measureable mass transfer of SiO_2 between wall rock and fluid in the upwelling zone of even vigorously flowing mid-ocean ridge hydrothermal system.

Cited Literature

Anbeek C (1992) Geochim Cosmochim Acta 56:3957–3970
Bischoff JL, Dickson FW (1975) Earth Planet Sci Lett 25:385–397
Blum AE, Yund RA, Lasaga AC (1990) Nature 331:431–433
Brady PV, Walther JV (1989) Geochim Cosmochim Acta 3:2823–22830

Brantley SL (1992) In: Kharaka YK, Maest AS (eds) Water–rock interaction. Balkema, Rotterdam, pp 3–6. ISBN 90 5610075
Brantley S (2004) In: Drever JI (ed) Treatise on geochemistry. Elsevier, Amsterdam, pp 73–118
Brantley SL, Crane SR, Crerar DA, Hellmann R, Stallard R (1986) Geochim Cosmochim Acta 50:2349–2361
Busenberg E, Clemency CV (1976) Geochim Cosmochim Acta 40:41–46
Busenberg E, Plummer LN (1996) U S Geolog Bull 1578:139–168
Casey WH (1987) J Geophys Res 92:8007–8013
Chase CG (1972) J Geophys Res 29:117–122
Christy AG, Putnis A (1993) Geochim Cosmochim Acta 57:2161–2168
Claasen HC, White AF (1979) In: Jenne EA (eds) Chemical modeling in aqueous systems, pp 771–793
Davies CW, Jones AL (1955) Trans Farady Soc 57:872–877
Dove PM, Crerar DA (1990) Geochim Cosmochim Acta 54:4147–4156
Dove PM, Platt FM (1996) Chem Geol 127:331–338
Fujii T (1977) In: Tatsumi T (ed) Basis of modern economic geology. University Tokyo Press, Tokyo (in Japanese)
Fujimoto K (1987) Min Geol 37:45–54 (in Japanese with English abstract)
Ganor J, Lasaga AC (1994) Min Mag 58A:315–316
Giggenbach WF (1984) Geochim Cosmochim Acta 48:2693–2711
Goldich SS (1938) J Geol 46:17–58
Grambow B (1985) Mat Res Soc Symp 44:15–27
Hannington MD, Petersen S, Jonasson IR, Franklin JM (1994) Geological survey of Canada open file report 2915C, 1 : 35000000 and CD-ROM
Holser WT, Kaplan IR (1966) Chem Geol 1:93–135
House WA (1981) J Chem Soc Faraday Trans I 77:341–359
Humphris SE, Thompson G (1978) Geochim Cosmochim Acta 42:107–125
Inskeep WP, Bloom PR (1985) Geochim Cosmochim Acta 49:2165–2180
Kazmierczak TF, Tomson MB, Nancollas GH (1982) Crystal growth of calcite carbonate. A controlled composition kinetic study. J Phys Chem 86:103–107
Kitahara S (1960) Rev Phys Chem Jpn 30:123–130
Lasaga AC (1980) Geochim Cosmochim Acta 44:815–828
Lasaga AC (1981a) In: Reviews in mineralogy (Am Min), vol 8, pp 69–110
Lasaga AC (1981b) In: Reviews in mineralogy (Am Min), vol 8, pp 135–170
Lasaga AC (1981c) In: Reviews in mineralogy (Am Min), vol 8, pp 261–320
Lasaga AC (1984) J Geophys Res 89:4009–4025
Lasaga AC (1989) Earth Planet Sci Lett 94:417–424
Lasaga AC, Holland HD (1976) Geochim Cosmochim Acta 40:257–266
Lasaga AC, Kirkpatrick RJ (eds) (1981) Reviews in mineralogy (Am Min), vol. 8
Lasaga A, Soler JM, Ganor J, Burch T, Nagy KL (1994) Geochim Cosmochim Acta 58:2361–2386
Li Y-H, Gregory S (1979) Geochim Cosmochim Acta 38:703–714
Mackenzie FT, Garrels RM (1966) Am J Sci 264:507–525
Meike A (1990) Geochim Cosmochim Acta 54:3347–3352
Mogollon JL, Perez DA, Monaco SL, Ganor J, Lasaga AC (1994) Min Mag 58A:619–620
Nagy KL, Lasaga AC (1992) Geochim Cosmochim Acta 56:3093–3111
Nancollas GH, Liu ST (1975) Soc Pet Eng J 15:509
Nancollas GH, Reddy MM (1971) J Colloid Interface Sci 37:824–830
Nerlentniaks I (1980) J Geophys Res 85:4379–4397
Nielsen AE (1958) Acta Chem Scand 12:951–958
Nielsen AE (1983) In: Kolthoff IM, Elving PJ (eds) Treatise on analytical chemistry. Wiley, New York, pp 268–347
Nielsen AE, Toft JM (1984) J Cryst Growth 67:278–288
Nishiyama T (1987) Monthly Earth 91:54–59 (in Japanese)

Plummer LN, Wigliey TML, Parkhurst DL (1979) In: Chemical modeling in aqueous systems. A C S Sym Ser No. 93. American Chemical Society, Washington DC, pp 539–573; Chap. 25
Reddy MM, Gaillard WD (1981) J Colloid Interface Sci 80:171–178
Reddy MM, Nancollas GH (1971) J Colloid Interface Sci 36:166–172
Rimstidt JD, Barnes HL (1980) Geochim Cosmochim Acta 44:1683–1699; Acta 54:955–969
Sato T (1977) In: Volcanic processes in ore genesis, vol 6. Elsevier, Amsterdam, pp 129–222
Schnoor JL (1990) In: Stumm W (ed) Aquatic chemical kinetics. Wiley, New York, pp 475–504
Schoonen MAA, Barnes HL (1991) Geochim Cosmochim Acta 55:1505–1514
Schott J, Pettit JC (1987) In: Stumm W (ed) Aquatic surface chemistry. Wiley, New York, pp 293–312
Shikazono N (1989) Chem Geol 76:239–247
Shikazono N, Fujimoto K (1996) Chikyukagaku (Geochem) 30:91–97 (in Japanese with English abstract)
Shikazono N, Fujimoto K (2001) Bull Earth Res Inst Univ Tokyo 76:333–340
Shikazono N, Shiraki R (1994) Res Geol 44:379–390 (in Japanese with English abstract)
Shiraki R, Brantley SL (1995) Geochim Cosmochim Acta 59:1457–1471
Skagius K, Nerentnieks I (1986) Water Resour Res 22:389–398
Sorai M, Sasaki M (2010) Am Min 95:853–862
Steefel CI, Cappellen PV (1990) Geochim Cosmochim Acta 54:2657–2677
Tester JW, Wopoley WG, Robinson BA, Grigsby CO, Feerer JL (1994) Geochim Cosmochim Acta 58:2407–2420
Tivey MK, McDuff RE (1990) J Geophys Res 95:12617–12637
Uchida T (1992) Res Geol 42:175–190 (in Japanese with English abstract)
Wells JT, Ghiorso MS (1991) Geochim Cosmochim Acta 55:2467–2481
White AF (2004) In: Drever JL (ed) Treatise on geochemistry, vol 5. Elsevier, Amsterdam, pp 133–168
Wood BJ, Walther JV (1983) Science 222:413–415

Further Reading

Berner RA (1971) Principles of chemical sedimentology. McGraw-Hill, New York
Berner RA (1980) Early diagenesis—a theoretical approach. Princeton University Press, Princeton
Bidoglio G, Stumm W (1994) Chemistry of aquatic systems, local and global perspectives. Kluwer, Dordrecht
Cussler EL (1984) Diffusion, mass transfer in fluid systems. Cambridge University Press, Cambridge
Denbigh KG, Turner JCR (1971) Chemical reactor theory. Cambridge University Press, London
Frost AA, Pearson RG (1953) Kinetics and mechanism. Wiley, New York
Garrels RM, Mackenzie FT (1971) Evolution of sedimentary rocks. W W Norton and Co Inc, New York
Gordon L, Salutsky MC, Willard HH (1959) Precipitation from homogeneous solution. Wiley, New York
Langmuir D (1997) Aqueous environmental geochemistry. Prentice-Hall, New Jersey
Lasaga AC (1998) Kinetic theory in the earth sciences. Princeton University Press, Princeton
Lerman A (1979) Geochemical processes water and sediment environments. Wiley, New York
Levenspiel O (1972) Chemical reaction engineering, 2nd edn. Wiley, New York
Lichtner PC, Steefel CI, Oelkers EH (eds) (1996) Reaction transport in porous media. Reviews in mineralogy (Am Min), vol 34
Margenau H, Murphy GM (1956) The mathematics of physics and chemistry. D van Nostrand, Princeton

More FMM (1983) Principles of aquatic chemistry. Wiley, New York
Morel FMM, Morgan JJ (1968) A numerical method of solution of chemical equilibria in aqueous systems. California Institute of Technology, Pasadena
Nakano M (1991) Mass transfer in soil. University Tokyo Press, Tokyo (in Japanese)
Nielsen AE (1964) Kinetics of precipitation. Pergamon, Oxford
Phillips OM (1991) Flow and reactions in permeable rocks. Cambridge University Press, Cambridge
Robinson RA, Stokes RH (1959) Electrolyte solutions. Academic, New York
Shikazono N (2003) Geochemical and tectonic evolution of arc-backarc hydrothermal systems. Elsevier, Amsterdam
Stumm W, Morgan JJ (1970) Aquatic chemistry. Wiley, New York
Stumm W, Morgan JJ (1981) Aquatic chemistry chemical equilibria and rates in natural waters. Wiley, New York
Takamatsu T, Naito M, Fan L-T (1977) Environmental system technology. Nikkan Kogyo Shinbunsya, Tokyo (in Japanese)
Tosaka H (2006) Geosphere environment fluid flows: theories, models and applications. University Tokyo Press, Tokyo (in Japanese)

Chapter 4
System Analysis

In previous Chapters, various physico-chemical interactions and mass transfer mechanisms are considered. Analysis of a given system such as hydrothermal system and seawater system will be carried out below because hydrothermal system and seawater system are considered as representative endogenic and exogenic system, respectively. However, it is shown in this Chapter that both systems (hydrothermal system and seawater system) are influenced both by endogenic and exogenic energies. Interaction between hydrothermal system and seawater system is also considered below.

4.1 Hydrothermal System

There are many types of hydrothermal system and associated hydrothermal ore deposits. Formation of hydrothermal ore deposits and mass transfer in seafloor hydrothermal system will be considered below.

Schematic representation of seafloor hydrothermal system is shown in Fig. 4.1. Seafloor hydrothermal system consists of recharge zone, reservoir (reaction zone), heat source and discharge zone. Seawater penetrates downward from recharge zone. Downward migrating seawater is heated by the heat from magma and becomes to hydrothermal solution interacting with surrounding rocks mainly in reservoir, leaching elements such as base metal elements (Cu, Zn, Pb, etc.) from the rocks. Hydrothermal solution ascends along cracks and vents from the sea-floor at discharge zone. Hydrothermal solution mixes with cold seawater rapidly above seafloor and under sub-seafloor environment, resulting to the precipitation of minerals and formation of mound and chimney (Fig. 4.2) (Appendix, Plate 25). Mass transfer in seafloor hydrothermal system accompanied by the formation of ore deposits is understood by dividing the system into several parts (recharge zone, reservoir, discharge zone) as shown in Fig. 4.2.

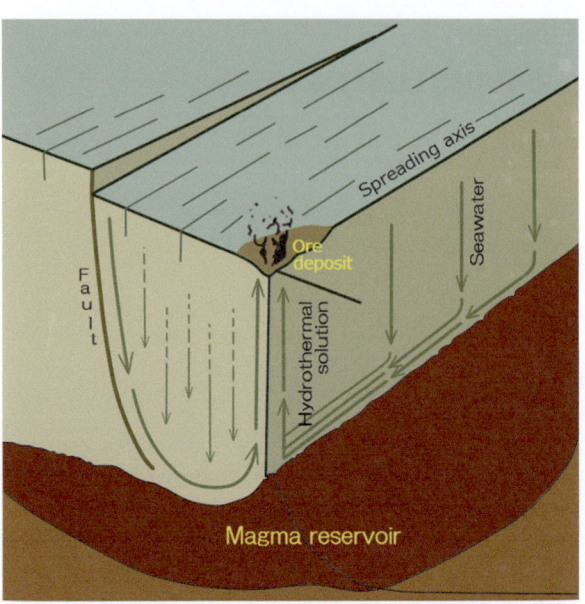

Fig. 4.1 Submarine hydrothermal system and formation of ore deposits (Shikazono 1992)

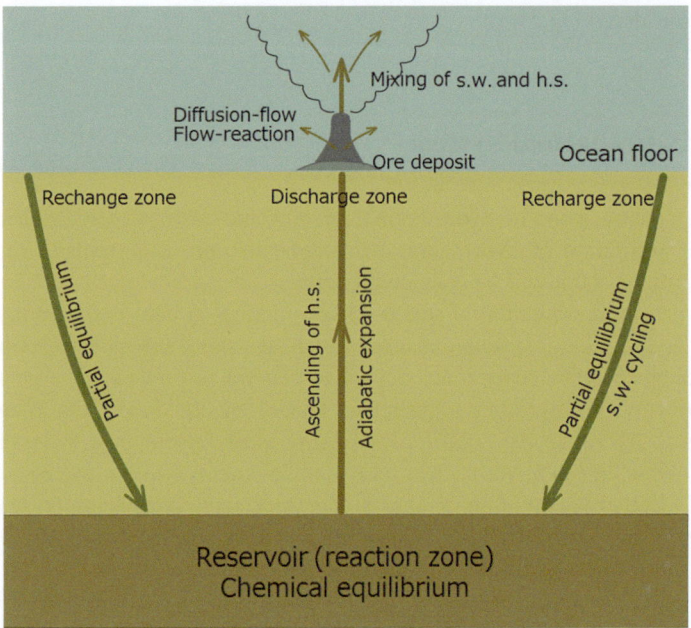

Fig. 4.2 Mass transfer in submarine hydrothermal system. *s.w.* seawater, *h.s* hydrothermal solution

Partial equilibrium model, chemical equilibrium model and adiabatic ascending model have been applied to recharge zone, reservoir and discharge zone, respectively in order to understand the geochemical features of hydrothermal solution and rocks. Several models to interpreting the formation of ore deposits on the seafloor have been proposed. For instance, mixing of hydrothermal solution with seawater, hydrothermal solution migrating through chimney shell, settling of mineral particle from hydrothermal solution mixed with seawater are interpreted in terms of partial chemical equilibrium, diffusion-flow, and settling-dispersion models, respectively. Rate-limiting mechanism (dissolution, precipitation, flow, diffusion, dispersion) is different for different subsystem.

4.1.1 Recharge Zone

Seawater penetrating downward from recharge zone interacts with basalt which is the main rock of upper part of oceanic crust. The concentration of base metal elements in hydrothermal solution of seawater origin increases with the proceeding of the interaction. The increase in the concentration of base metal elements in hydrothermal solution with an increase in temperature is caused by more stable existence of base-metal chloro-complexes at higher temperature. The concentration of H_2S in modified seawater (hydrothermal solution) increases with increasing temperature. H_2S in hydrothermal solution comes from reduction of seawater SO_4^{2-} and leaching from basalt. Sulfur isotope study revealed that the contribution of sulfide sulfur of basalt origin is larger than that of marine origin (Kawahata and Shikazono 1988). $\delta^{34}S$ of hydrothermal solution is +2 to +8 ‰ that lies between seawater sulfate sulfur value (+20 ‰) and basalt sulfide sulfur value (+1 ‰), but is closer to the basaltic sulfur value, indicating larger contribution of basalt sulfide sulfur than seawater sulfate sulfur. $\delta^{34}S$ of hydrothermal solution is determined by the following mass balance equation.

$$m_{iR}\delta^{34}S_{iR}W_R + m_{iS}\delta^{34}S_{iS}W_S = m_{eqR}\delta^{34}S_{eqR}W_R + m_{eqS}\delta^{34}S_{eqS}W_S \quad (4.1)$$

where R is rock, S is seawater, i is initial state, $eq.$ is equilibrium state, m is concentration of sulfur and W is weight.

Concentrations of elements such as alkali and alkali earth elements change by seawater–rock interaction (Shikazono 1988). The concentration of ore-forming elements (Ba, Zn and other base metals) increase with increasing temperature. The increase in the concentration of K and Ba is probably due to the decomposition of feldspar and volcanic glass. The concentration of Mg in hydrothermal solution decreases with increasing temperature which is due to the formation of Mg-silicates (e.g., Mg-smectite and Mg-chlorite) at elevated temperatures. The concentration of Na in hydrothermal solution does not change, irrespective of the formation of Na-feldspar. The concentration of Ca decreases due to the precipitation of anhydrite ($CaSO_4$) at initial stage of seawater–rock interaction, but it increases probably due

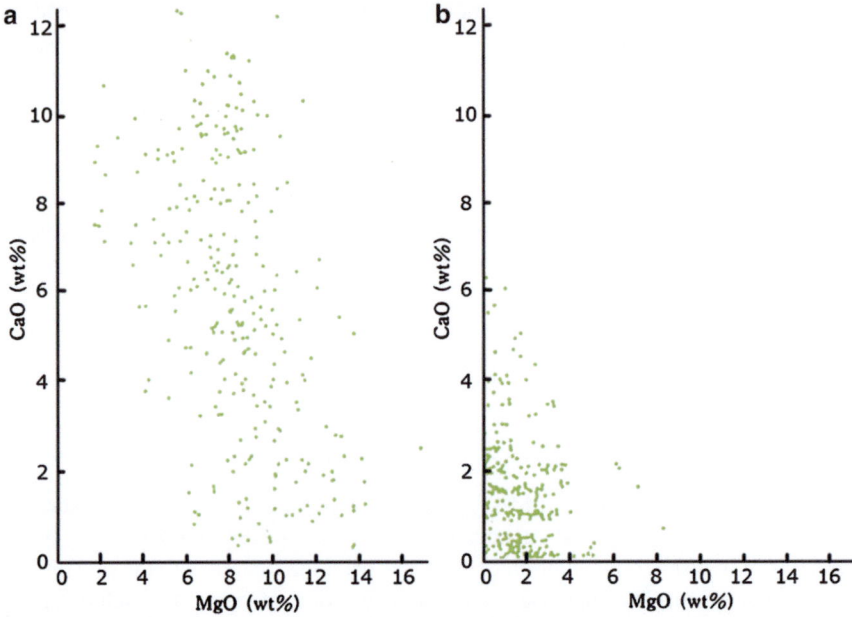

Fig. 4.3 Relationship between MgO content and CaO content in altered volcanic rocks in Green tuff region (near Kuroko mining area) (Shikazono 1994a, b). (**a**) basalt, (**b**) dacite

to the exchange reaction of Mg in seawater (or modified seawater) for Ca in rocks. The concentration of Sr in hydrothermal solution is relatively constant.

$^{87}Sr/^{86}Sr$ of hydrothermal solution is 0.703 which lies between basalt value (0.701) and seawater value (0.709). This indicates that most of Sr in hydrothermal solution is of basalt origin, but it also comes from seawater. Sr of seawater origin moves to anhydrite formed at initial stage. It comes to hydrothermal solution by the decompositions of silicates (feldspar, volcanic glass etc.).

The relationship between MgO content and CaO content of altered igneous rocks (basalt and dacite) from Green tuff region (Tertiary submarine thick volcanic pile which is intensely altered mainly by hydrothermal activity and diagenesis and is greenish in appearance due to the presence of clay minerals (smectite, chlorite, celadonite) in Japan and other Circum Pacific region) is shown in Fig. 4.3 (Shikazono 1994a, b).

Inverse correlation of MgO content with CaO content in altered igneous rocks is found (Fig. 4.3). This is caused by the reaction

$$Mg^{2+}(\text{seawater}) + CaO\,(\text{rock})$$
$$\rightarrow Ca^{2+} + MgO\,(\text{alteration minerals; Mg-smectite, Mg-chlorite}) \quad (4.2)$$

Therefore, MgO content of altered rocks indicates the extent of seawater–rock interaction (degree of alteration). The contents of some elements (Fe, H_2O) positively correlate with MgO content, but some (CaO, K_2O) negatively do.

4.1 Hydrothermal System

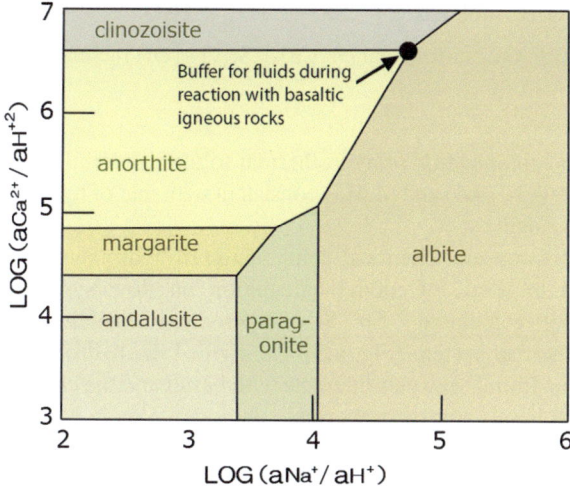

Fig. 4.4 Phase diagram for CaO–Na$_2$O–SiO$_2$–Al$_2$O$_3$–H$_2$O system in equilibrium with quartz at 400 °C and 400 bars. Plagioclase solid solution can be represented by the albite and anorthite fields, whereas epidote is represented by clinozoisite. Note that the clinozoisite field is adjacent to the anorthite field, suggesting that fields with high Ca/[H$^+$]^2 might equilibrate with excess anorthite by replacing it with epidote. The location of the albite-anorthite-epidote equilibrium point is a function of epidote and plagioclase composition and depends on the model used for calculation of the thermodynamic properties of aqueous cations (Berndt et al. 1989)

The decrease in SiO$_2$ content with increasing MgO content is due to the formation of Mg-chlorite containing smaller SiO$_2$ content than original rock.

4.1.2 Reservoir

Chemical compositions of hydrothermal solution at mid-oceanic ridges are uniform, suggesting the attainment of chemical equilibrium between alteration minerals and hydrothermal solution in reservoir. It is shown in Fig. 4.4 that chemical composition of hydrothermal solution venting from the ridge axis is buffered by epidote-feldspar assemblage (Berndt et al. 1989). It can be derived from the following reactions that $a_{Ca^{2+}}/a_{H^{+2}}$ and a_{Na^+}/a_{H^+} ratios of hydrothermal solution relate to temperature, activity of clinozoisite component in epidote, and activity of Ca-feldspar component in feldspar.

$$CaAl_2Si_2O_8(\text{Ca-feldspar}) + 2Na^+ + 4SiO_2$$
$$= 2NaAlSi_3O_8(\text{Na-feldspar}) + Ca^{2+} \qquad (4.3)$$

$$3CaAl_2Si_2O_8(\text{Ca-feldspar}) + Ca^{2+} + 2H_2O$$
$$= 2Ca_2Al_3Si_3O_{12}(OH)(\text{clinozoisite}) + 2H^+ \qquad (4.4)$$

$$2CaAl_2Si_2O_8(\text{Ca-feldspar}) + 2SiO_2 + Na^+ + H_2O$$
$$= NaAlSi_3O_8(\text{Na-feldspar}) + CaAl_3Si_3O_{12}(OH)(\text{clinozoisite}) + 2H^+$$
(4.5)

The chemical composition of hydrothermal solution determined by the chemical equilibrium for (4.3), (4.4) and (4.5) is consistent with that of hydrothermal solution at mid-oceanic ridge.

The relatively constant chemical composition of hydrothermal solution can be also explained in terms of model calculation on flow system. For example, Kawahata (1989) calculated $^{87}Sr/^{86}Sr$ of hydrothermal solution passing through cells of hydrothermal system (Fig. 4.5). He divided the hydrothermal system into 100 and 50 cells. Initial seawater interacts with basalt and the chemical equilibrium between altered basalt and hydrothermal solution attains in a first cell. Then, the solution moves to the second cell and interacts with basalt and the equilibrium between basalt and solution attains. The stepwise interaction occurs in the cells with $^{87}Sr/^{86}Sr$ of initial seawater $= 0.709$, but it decreases by the interaction of basalt with $^{87}Sr/^{86}Sr = 0.702$. According to this calculation, $^{87}Sr/^{86}Sr$ of hydrothermal solution in reservoir and discharge zone is constant (0.7035) which is consistent with that of hydrothermal solution discharging at mid-oceanic ridge.

4.1.3 Discharge Zone

If fracturing occurs in hydrothermal system, hydrothermal solution in reservoir begins to ascend very rapidly along the fractures. It seems likely as an approximation that hydrothermal solution ascends with adiabatic expansion without the heat loss from hydrothermal solution to surrounding rocks because of very rapid flow rate of hydrothermal solution (1–10 m s^{-1}) at the sea floor from which hydrothermal solution issues. Temperature of hydrothermal solution decreases due to adiabatic expansion. Hydrothermal solution sometimes boils by the changes in temperature and pressure in underground.

Vapor and gaseous components release from hydrothermal solution accompanied by the boiling, resulting to higher salinity of hydrothermal solution. It is considered that fluid inclusions with high salinity and NaCl crystal in gabbro from oceanic crust are caused by the boiling of hydrothermal solution. Sulfide minerals sometimes form precipitate due to the changes in temperature and pressure of ascending hydrothermal solution. However, mostly hydrothermal solution issues from the seafloor with no precipitation of minerals during the ascent of hydrothermal solution. Ascending hydrothermal solution leaches SiO_2 from the country rocks (Wells and Ghiorso 1991).

The precipitated amount of quartz from the ascending hydrothermal solution is calculated based on the following equation.

4.1 Hydrothermal System

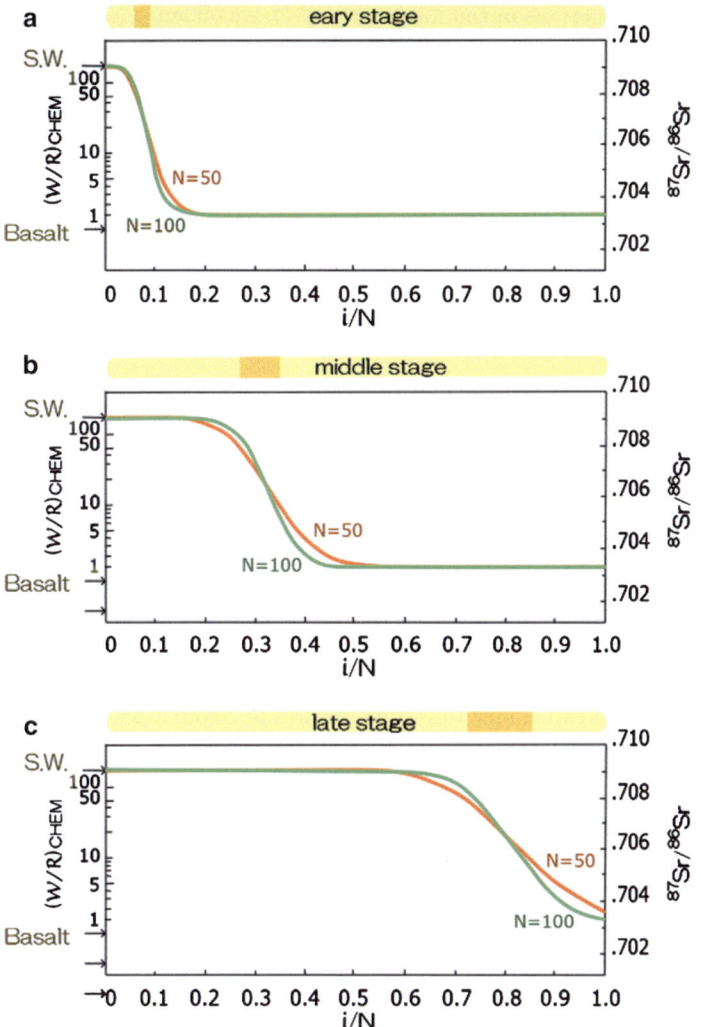

Fig. 4.5 Variation of $^{87}Sr/^{86}Sr$ as a function of i/N (Kawahata 1989). i: ith cell, N: total number of all. (**a**) early stage ($[W/R]^{SYS}_{FLOW} = 1$), (**b**) middle stage ($[W/R]^{SYS}_{FLOW} = 4$), (**c**) late stage ($[W/R]^{SYS}_{FLOW} = 10$), s.w. seawater

$$dm/dt = -k(A/M)a_{H_4SiO_4} + vdm/dx \tag{4.6}$$

where m is H_4SiO_4 concentration, k is dissolution rate constant, $a_{H_4SiO_4}$ is activity of H_4SiO_4, v is velocity of hydrothermal solution, and A/M is surface area of minerals/mass of fluid ratio.

At steady state equation (4.6) converts into

$$-k(A/M)a_{H_4SiO_4} + vdm/dx = 0 \tag{4.7}$$

Expressing $a_{H_4SiO_4}$ as m and integrating (4.7), we obtain

$$lnm = lnm_i - k(A/Mv)x \qquad (4.8)$$

where m_i is initial concentration.

If $k = 10^{-6}$ (at 200 °C), $A/M = 0.2$ (which corresponds to 1 cm width, and $v = 10^{-4}$ m s, we obtain $m = m_i$. This implies no precipitation of SiO_2 during the ascent of hydrothermal solution. Equation (4.8) shows that it is more difficult to form quartz under the conditions of higher velocity, lower A/M and lower k at lower temperature.

4.1.4 Formation of Chimney and Sea-Floor Hydrothermal Ore Deposits and Precipitation Kinetics-Fluid Flow Model

Chimney (Fig. 4.2) is mainly composed of sulfides, sulfates and silica which precipitated by the mixing of hydrothermal solution and cold ambient seawater. Formation mechanism of chimney has been explained in terms of several models. Partial equilibrium model was applied to the precipitations of minerals during the mixing of hydrothermal solution and cold seawater (Ohmoto et al. 1983). However, this model cannot explain the following features of chimney and submarine hydrothermal ore deposits.

(1) Metastable phases such as wurtzite, amorphous silica and native sulfur are common.
(2) Positive correlation between the amount of quartz and barite, that is predicted based on equilibrium model, is not found: Solubility decreases with decreasing temperature, indicating that quartz precipitates due to the mixing. Solubility of barite in the solution with more than 1 mol·kg^{-1} H$_2$O NaCl concentration decreases with decreasing temperature. Barite precipitates due to the increasing of SO_4^{2-} concentration during the mixing of hydrothermal solution and seawater accompanied by the decreasing of temperature. This indicates that the precipitated amount of barite increases with decreasing temperature. Therefore, it is inferred that the amount of quartz precipitated positively correlates with that of barite precipitated. However, this correlation is not found in the Kuroko ore.
(3) The ratio of sulfide content/sulfate content of ore cannot be explained by equilibrium model of seawater-hydrothermal solution mixing.
(4) Equilibrium is not attained for the partitioning of Sr between anhydrite and mixed fluid of hydrothermal solution and seawater (Fig. 4.6).

Above evidence indicates that the mixing of hydrothermal solution and seawater is very rapid, resulting to rapid precipitation of minerals under the conditions far from equilibrium.

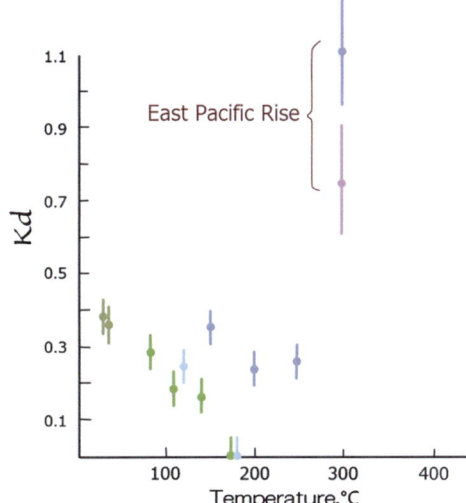

Fig. 4.6 Partition coefficient of Sr (Kd) (Shikazono and Holland 1983)

Shikazono et al. (2012) used precipitation kinetics-fluid flow model to explain the mode of occurrences of barite, quartz and anhydrite in Kuroko deposits (submarine hydrothermal polymetallic sulfide–sulfate massive deposits in Japan).

They calculated the amounts of quartz and barite precipitated from hydrothermal solution mixed with seawater based on the equation for quartz

$$dm/dt = -k(A/M)(m - m_{eq}) + (q/V)(m_i - m) \quad (4.9)$$

where m is H_4SiO_4 concentration, m_{eq} is equilibrium H_4SiO_4 concentration, m_i is initial H_4SiO_4 concentration prior to the mixing of hydrothermal solution and cold seawater, q is volume flow rate, and v is volume of the system.

For Barite

$$dm_{Ba^{2+}}/dt = -k(A/M)\left\{(m_{Ba^{2+}}m_{SO_{4i}})^{1/2} - (K_{sp}/\gamma^2)^{1/2}\right\}^2 + q/V(m_{Ba_i} - m_{Ba^{2+}}) \quad (4.10)$$

where m_{Ba} is Ba concentration, m_{Bai} is initial Ba concentration, $m_{SO_{4i}}$ is initial SO_4^{2-} concentration, γ is average activity coefficient of Ba^{2+} and SO_4^{2-} and K_{sp} is solubility product of $BaSO_4$.

The calculated results showing the relationships between precipitated amounts of barite and quartz and velocity and A/M are shown in Figs. 4.7 and 4.8.

Those indicate that the conditions favorable for the precipitations of quartz and barite are distinctly different: Quartz can precipitate from the solution having relatively high temperature and low flow rate and under high A/M, while barite can precipitate from the solution with relatively high flow rate and low temperature and under low A/M compared with quartz precipitation.

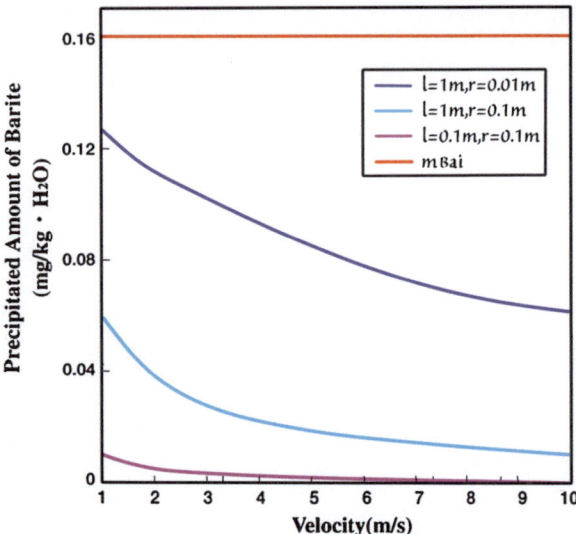

Fig. 4.7 Relationship between precipitated amount of barite and velocity of fluid at 200 °C. v velocity (m/s), m_{Bai} initial concentration of Ba

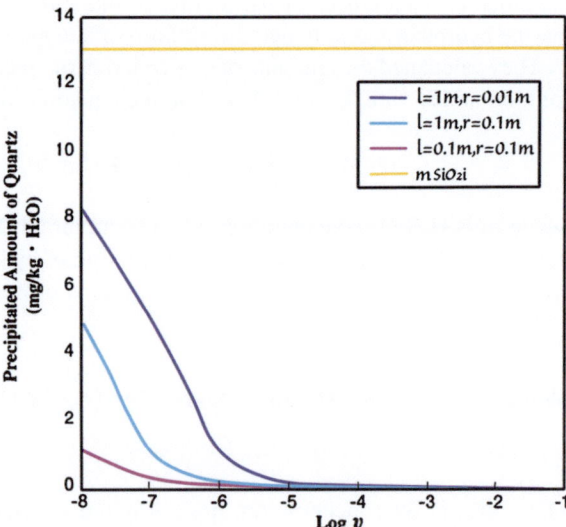

Fig. 4.8 Relationship between precipitated amount of quartz and velocity of fluid at 200 °C. m_{SiO_2i}, initial concentration of SiO_2, v velocity (m/s)

The results of the calculations are in agreement with the occurrences of barite and quartz and chemical features of discharging fluids in the submarine hydrothermal system (Shikazono et al. 2012).

Concentration in aqueous solution is homogeneous in perfectly mixing model. However, in fact, concentration varies in a system. According to piston flow model the variation in concentration is expressed as

$$m = m_i \exp(-kx/v) \tag{4.11}$$

where m_i is initial concentration of input solution, k is precipitation rate constant, and v is velocity of solution.

Concentration, residence time and efficiency of precipitation vary with different mode of flow.

Shikazono and Holland (1983) and Shikazono (1994a, 1994b) have shown experimentally that morphology of anhydrite and barite precipitating from hydrothermal solution varies with the degree of supersaturation; dendritic (rod-like, spindle-like, star-like, cross-like) crystals with rough surface, where the crystal growth is controlled by the diffusional species, formed from aqueous solution with high degree of supersaturation while well-formed rectangular, rhombohedral, and polyhedral crystals with smooth surfaces, where a surface reaction predominates, precipitated from low degrees of supersaturation. Therefore, it is likely that rate constant values for anhydrite and barite used by Shikazono et al. (2012) should be different from those in natural seafloor hydrothermal system associated with Kuroko mineralization which takes place under supersaturated condition due to the rapid mixing of hydrothermal solution and seawater and thus the calculations on the precipitated amounts of anhydrite and barite should be done in future, considering the effect of degree of supersaturation.

4.1.5 Precipitation–Dispersion Model

Very fine mineral particles precipitate by a rapid mixing of hydrothermal solution with seawater and transport laterally, affected by ocean current. Vertical settling of mineral particle from the plume is controlled by Stokes equation

$$v_s = 2\alpha g(\rho_s - \rho)r^2/9\mu \tag{4.12}$$

where v_s is settling velocity, α is configuration factor, ρ is density, r is particle size, μ is viscosity, and s is mineral particle.

The calculations on the settling velocity using (4.12) indicate that the particles with more than 100 μm in diameter settle from the plume onto seafloor, but the particles with less than 100 μm do not. Electron microscopic observation of the chimneys from Kuroko and mid-oceanic ridge deposits supports this calculation results.

Above consideration concerns only with the vertical settling of mineral particles. Feely et al. (1987) considered the effect of dispersion of black smoker mineral particles from active vents on the Juan de Fuca ridge in addition to the settling and obtained the distribution of the size of mineral particles settled onto the seafloor from the equation

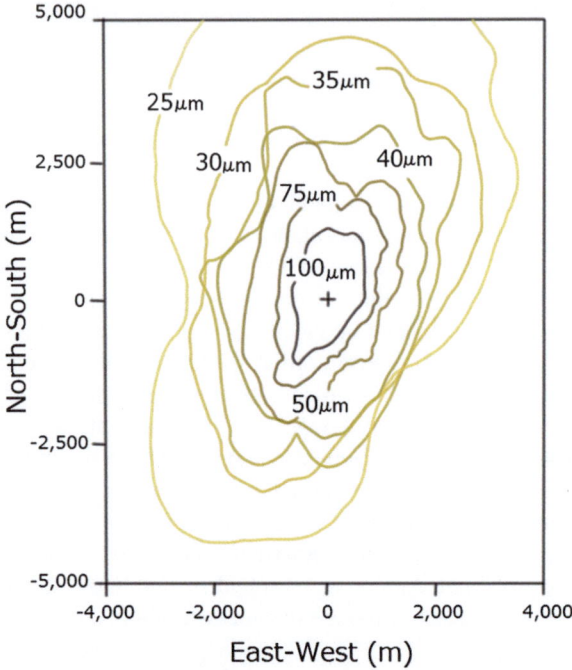

Fig. 4.9 Distribution pattern of precipitated pyrite grain size around hydrothermal vent at south Juan de Fucca ridge (Feeley et al. 1987)

$$\partial m/\partial t + u(t)\partial m/\partial x + v(t)\partial m/\partial y + A_H\left(\partial^2 m/\partial x^2 + \partial^2 m/\partial y^2\right)$$
$$- w_s(t)\partial m/\partial z = 0 \qquad (4.13)$$

where m is concentration of particle, $u(t)$ is velocity in x direction, $v(t)$ is velocity in y direction, $w_s(t)$ is velocity in z direction, and A_H is eddy diffusion coefficient.

The particles are assumed to have sufficiently large settling velocity that diffusion in the vertical direction has little effect on overall particle movement. The results of calculations clarified that the pyrite particles with small grain size less than 100 μm do not settle near the vent of hydrothermal solution but the majority of large-grained black smoker particles should be deposited within a few hundred meters of the vent (Fig. 4.9).

4.1.6 Diffusion-Fluid Flow Model

Silica coexists with barite with large grain size in chimney. The grain size of silica is very small, indicating that it does not settle from plume onto seafloor near the vent and disperse laterally. Figure 4.10 shows the temperature variation during mixing of hydrothermal solution and seawater and conductive cooling. This indicates that silica precipitates due to the cooling, but not due to the mixing

4.1 Hydrothermal System

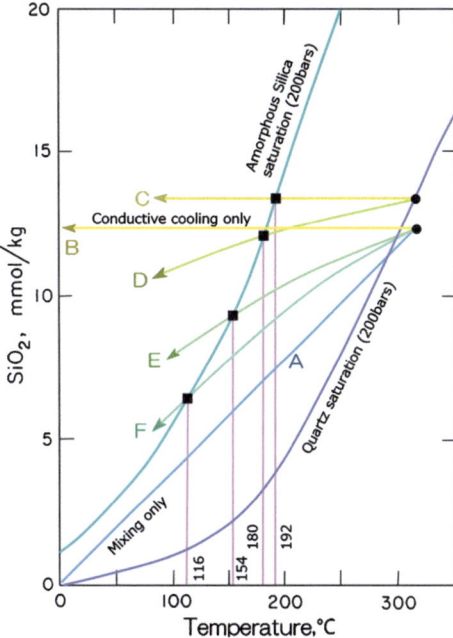

Fig. 4.10 Conductive cooling combined with mixing to cause the precipitation of amorphous silica in chimneys from two different fluids at Guaymas Basin (Peter and Scott 1988). As in Figure, quartz does not precipitate for kinetic reasons nor does amorphous silica by simple mixing of the vent fluid with ambient seawater (line *A*). Amorphous silica does precipitate as a consequence of conductive cooling and mixing. The relative amounts of mixing may be determined accurately from the intercepts of the fluid inclusion homogenization temperatures with the silica saturation curve (lines *B–F*) (Scott 1997)

(Herzig et al. 1988). Silica diffuses in hydrothermal solution migrating in chimney interior and precipitates due to the cooling by heat loss. Silica precipitation in chimney interior is explained by this diffusion-flow model. Tively and Mcduff (1990) explained mineral zoning in chimney based on diffusion-flow model, using diffusion coefficient of dissolved species in water that was estimated from porosity and tortuosity. From the profile of concentration of dissolved species in hydrothermal solution in chimney the site of the precipitation of mineral can be known. Flow rate is deduced from Darcy's law. Permeability is given for the chimney sample. Hydraulic gradient is obtained from the relation.

$$P_i = P_r + (\rho_i/2)(W_r^2 - W_i^2) + (\rho_i - \rho_0)g(Z_r - Z_i) \\ + \rho_i \sum (1/2)(W^2 Lf/R_h)_i + \rho_i \sum (1/2)(W^2 e_\nu)_i \quad (4.14)$$

where P is pressure, ρ is density of fluid, f is fanning factor, R_h is cross section area/length of circumference ratio, W is flow rate in pipe, Z is depth, e_ν is friction loss factor, r is upper surface of pipe, and i is inner part of pipe. The pressure gradient was estimated from the fluid dynamics of high-velocity fluid flow inside the chimney. The parametric studies revealed the importance of diffusion driving gradients in activity coefficients resulting from the strong thermal gradient.

4.1.7 Dissolution-Recrystallization Model

Ascending hydrothermal solution interacts with ore body, resulting to dissolution of minerals and recrystallization.

Very fine-grained minerals precipitate from oversaturated mixed fluid of hydrothermal solution and seawater. Solubility of fine-grained mineral is generally higher than large-grained mineral. Fine-grained mineral changes to large-grained mineral. This process is called Ostwald ripening.

The following Gibbs-Kelvin equation is established for the solution saturated with grain with size r* (Nielsen 1964) which is given by

$$r* = 2\sigma(V_0/RT)ln(m/m_{eq}) \qquad (4.15)$$

where σ is surface free energy, V_0 is molar volume of mineral, m is concentration in aqueous solution and m_{eq} is bulk solubility (solubility of very large grain).

From (4.15) the ratio of solubility of grain with size r (m_r) to bulk concentration (m) is

$$m_r/m = \exp\{2\sigma(V_0/RT)(1/r - 1/r*)\} \qquad (4.16)$$

This means that driving force for Ostwald ripening is surface free energy of mineral. Surface free energy of main mineral is listed in Table 4.1. Grain size of minerals changes during mixing of hydrothermal solution with seawater and by Ostwald ripening at post-depositional stage. The processes cause variable grain sizes of minerals.

Grain size depends on temperature and degree of supersaturation. It decreases with increasing degree of supersaturation. Grain size of ore constituent mineral in upper parts of Kuroko ore deposits is smaller. This is interpreted as that minerals with small grain size precipitated and settled onto the ore body on the sea bottom and the grain size of minerals in the interior of ore body became larger by Ostwald ripening. The change of the grain size and morphology of ore and gangue minerals (e.g., sulfides, sulfates) is observed in volcanogenic massive deposits (Kuroko-type deposits). In such recrystallization process, chemical composition of the minerals changes as well as grain size and morphology. Ogawa et al. (2007) analyzed Sr, Ba and rare earth elements (REE) and $^{87}Sr/^{86}Sr$ of anhydrite from Kuroko deposits in Japan and showed these contents decreased due to the recrystallization of fine-grained anhydrite to large-grained one based on the calculation of Sr content and $^{87}Sr/^{86}Sr$ of anhydrite during the precipitation due to the mixing of hydrothermal solution and Sr, REE contents, $^{87}Sr/^{86}Sr$ and microscopic observation.

4.1 Hydrothermal System

Table 4.1 Surface free energy of mineral and water (Lasaga 1997)

Mineral	Chemical formula	δ (mJ/m^2)
Fluorite	CaF$_2$	120
Calcite	CaCO$_3$	97
Gypsum	CaSO$_4$·2H$_2$O	26
Barite	BaSO$_4$	135
F-apatite	Ca$_5$(PO$_4$)$_3$F	289
OH-apatite	Ca$_5$(PO$_4$)$_3$OH	87
Portlandite	Ca(OH)$_2$	66
Brucite	Mg(OH)$_2$	123
Goethite	FeOOH	1,600
Hematite	Fe$_2$O$_3$	1,200
Gibbsite (001)	Al(OH)$_3$	140
Gibbsite (100)		483
Quartz	SiO$_2$	350
Amorphous silica	SiO$_2$	46
Kaolinite	Al$_2$Si$_2$O$_5$(OH)$_4$	>200
Water-air	H$_2$O	71.96
Pyrite (100)	FeS$_2$	3,155
Pyrite (111)	FeS$_2$	4,733

Grain size depends also on solubility as well as temperature and degree of supersaturation. Generally, grain size of minerals (e.g., sulfates) whose solubility is higher than those of low-solubility minerals (sulfides, quartz) is larger than the low-solubility minerals. With the proceeding of mixing of hydrothermal solution and seawater, degree of supersaturation for the minerals whose solubility decreases with decreasing temperature may become higher, if precipitation does not occur. If nucleation rate is slow even if degree of supersaturation is high, nucleation does not occur. Nucleation rate is important for controlling grain size. It is expressed as

$$dN/dt = J = J_o \exp(-\Delta G^*/k_b T) \quad (4.17)$$

where $\Delta G^* = 16\pi\sigma^3 V^2/(k_b T \ln\Omega^2)$, N is number of nuclei, V is volume, J_o is constant, and k_b is constant. This equation indicates that nucleation rate is slow when degree of supersaturation (Ω) is high.

4.1.8 Formation of Metastable Phase

Metastable phases such as amorphous silica (SiO$_2$) and wurtzite (ZnS) are common in mid-oceanic ridge deposits. These phases form from highly supersaturated mixed fluid of hydrothermal solution and seawater. Solubility of metastable phases is higher than that of stable phases. Therefore, metastable phases dissolve and stable phases form after the formation of ore deposits. Wurtzite (metastable phase of ZnS) is observed in active chimney, but sphalerite (stable phase of ZnS) is in nonactive chimney. It is inferred that sphalerite formed by the dissolution of wurtzite.

Anhydrite is abundant in active chimney but not inactive one. Anhydrite dissolves by the interaction of cold seawater because solubility of anhydrite at lower temperature is higher than that at higher temperature.

As mentioned above, metastable phases, fine-grained mineral particles and high solubility minerals dissolve and change to stable phases and large-grained crystals at post-depositional stages. Characteristic features of submarine hydrothermal ore deposits can be interpreted in terms of dynamics and kinetics models. For example, vertical zoning of Kuroko ore body in ascending order is gypsum zone, siliceous ore zone, yellow ore zone, black ore zone, barite ore zone and ferruginous chert ore zone. This zoning cannot be explained in terms of chemical equilibrium model, but can be by dynamics and kinetics model (Shikazono et al. 2012).

4.2 Seawater System

4.2.1 Chemical Equilibrium and Steady State

Seawater chemistry is determined by several factors. River water input, hydrothermal solution input, evaporation of seawater, interaction of seawater with oceanic crust and anthropogenic influence are important factors. It is simply assumed that seawater chemistry is controlled by input and output fluxes and chemical reaction in the system. Reversible chemical reaction will be considered below.

$$A \underset{k'}{\overset{k}{\rightleftarrows}} B \qquad (4.18)$$

where k is rate constant for the reaction, $A \rightarrow B$, and k' is rate constant for the reaction, $B \rightarrow A$. These reactions occur in a perfectly mixing system with volume, V (Fig. 4.11). The solution with m_{A0} (initial concentration of A) and m_{B0}

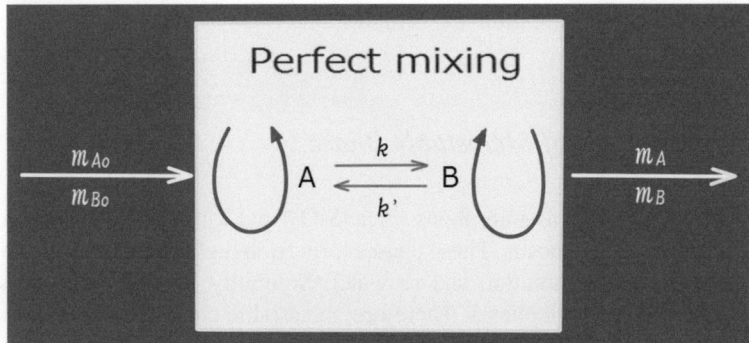

Fig. 4.11 Reversible reaction in perfectly mixing system and input and output of solution. m_{A0} and m_{B0} are input concentrations of A and B, respectively. m_A and m_B are output concentration of A and B, respectively

4.2 Seawater System

(initial concentration of B) inputs into the system. The solution with m_A and m_B outputs from the system. Volume flow rate of input solution and output solution is q. Mass balance equations for A and B are expressed as

$$dm_A/dt = (q/V)m_{A0} - (q/V)m_A - km_A + k'm_B \quad (4.19)$$

$$dm_B/dt = (q/V)m_{B0} - (q/V)m_B + km_A - k'm_B \quad (4.20)$$

If steady state is attained, that is, $dm_A/dt = dm_B/dt = 0$, the following equation can be derived.

$$m_B/m_A = \{k(V/q)(m_{A0} + m_{B0}) + m_{B0}\}/\{k'(V/q)(m_{A0} + m_{B0}) + m_{A0}\} \quad (4.21)$$

The relationship between equilibrium constant, $K_{4\text{-}18}$, and reaction rate constant, k and k', is

$$K_{4\text{-}18} = k/k' \quad (4.22)$$

Therefore

$$m_A/m_B = 1/K_{4\text{-}18} + 1/\{k(V/q)\} \quad (4.23)$$

$1/\{k(V/q)\}$ in (4.23) is a deviation from equilibrium. If residence time ($\tau = V/q$) is large, m_A/m_B is approximately equal to $1/K_{4\text{-}18}$, that means the system is close to equilibrium state. V is very large and q is small for seawater, compared with other natural waters (e.g., river water, lake water, ground water). Therefore, seawater system is tends to be close to equilibrium. On the contrary chemical equilibrium is usually not attained for terrestrial low temperature waters. The degree of attainment of chemical equilibrium depends on chemical composition and is considerably different for different element.

4.2.2 Chemical Equilibrium Model

Chemical equilibrium between seawater and marine sediments will be considered below following Sillen (1961, 1967a, b), Kramer (1965), Stumm and Morgan (1970), and Sayles and Margelsdorf (1977). Constituent minerals of sediments involve silicates, carbonates, phosphates and sulfates as follows.

Na-montmorillonite, H-montmorillonite, K-illite, H-illite, phillipsite ((K, Na, Ca)$_{2\text{-}4}$(Al,Si)$_{16}$O$_{32}$12H$_2$O), strontianite (SrCO$_3$), celestite (SrSO$_4$), chlorite, kaolinite, OH-apatite, FCO$_2$-apatite, calcite, aragonite, gypsum, and Mg-illite.

Chemical composition of seawater in equilibrium with above minerals can be calculated using equilibrium constants listed in Table 4.2, mass balance equations and electroneutrality relation. The calculated results by Kramer (1965) are mostly

Table 4.2 Chemical equilibrium and thermodynamic data used by Kramer (1965)

Reaction	Equilibrium constant
H-montmorillonite (C site) = Na-montmorillonite (C site)	$10^{-3.2} = \dfrac{[NaC](H^+)}{[HC](Na^+)}$
H-montmorillonite (E site) = Na-montmorillonite (E site)	$10^{-7.4} = \dfrac{[NaE](H)}{[HE](Na)}$
H-illite (C site) = K-illite (C site)	$10^{-2.4} = \dfrac{[KC](H)}{[HC](k)}$
H-illite (E site) = K-illite (E site)	$10^{-5.7} = \dfrac{[KE](H)}{[HE](K)}$
$CaSO_4 \cdot 2H_2O = Ca^{2+} + SO_4^{2-} + 2H_2O$	$10^{-4.60} = (Ca^{2+})(SO_4^{2-})$
Ca^{2+} + H-illite = Ca-illite + $2H^+$	$10^{-3.8} = (Ca^{2+})(H^+)^{0.08}$
$Ca_2Al_4Si_3O_{24} \cdot 9H_2O$(phillipsite) + $4H^+$ = $4SiO_2$(quartz) + $2Al_2Si_2O_7 \cdot 2H_2O$(kaolinite) + $2Ca^{2+}$ + $7H_2O$	$10^{13} = (Ca^{2+})/(H^+)^2$
Mg^{2+} + H-illite = Mg-illite + $2H^+$	$10^{-4.4} = (Mg^{2+})/(H^+)^{-0.1}$
$Mg_5Al_2SiO_3O_{14} \cdot 4H_2O$(chlorite) + $10H^+$ = SiO_2(quartz) + $Al_2SiO_2 \cdot 2H_2O$(kaolinite) + $7H_2O$ + $5Mg^{2+}$	$10^{14.2} = (Mg^{2+})/(H^+)^2$
$Ca_{10}(PO_4)_4(OH)_2 = 10Ca^{2+} + 6PO_4^{3-} + 2OH^-$	$10^{-112} = (Ca^{2+})^{10}(PO_4^{3-})^6(OH^-)^2$
$[Na_{0.286}Ca_{9.56}][(PO_4)_{0.35}(CO_3)_{0.33}][F_{2.04}]$	Equilibrium constant = 10^{-103}
$CaCO_3$(calcite) = $Ca^{2+} + CO_3^{2-}$	$10^{-8.09} = (Ca^{2+})(CO_3^{2-})$
$CaCO_3$(aragonite) = $Ca^{2+} + CO_2^{2-}$	$10^{-7.92} = (Ca^{2+})(CO_3^{2-})$
$CaMg(CO_3)_2 = Ca^{2+} + Mg^{2+} + 2CO_3^{2-}$	$10^{-18.21} = (Ca^{2+})(Mg^{2+})(CO_3^{2-})^2$
$MgCO_3 \cdot 3H_2O = Mg^{2+} + CO_3^{2-} + 3H_2O$	$10^{-5} = (Mg^{2+})(CO_3^{2-})$
$H_2CO_3 = H^+ + HCO_3^-$	$10^{-0.32} = (H^+)(HCO_3^-)/(H_2CO_3)$
$HCO_3^- = H^+ + CO_3^{2-}$	$10^{-10.6} = (H^+)(CO_3^{2-})/(HCO_3^-)$
$CO_2(g) + H_2O = H_2CO_3$	$10^{+1.19} = P_{CO_2}/(H_2CO_3)$
$CaHCO_3^+ = Ca^{2+} + HCO_3^-$	$10^{-1.28} = (Ca^{2+})(HCO_3^-)/(CaHCO_3^+)$
$NaCO_3^- = Na^+ + HCO_3^-$	$10^{-1.27} = (Na^+)(HCO_3^-)/(NaCO_3^-)$
$CaCO_3^0 = Ca^{2+} + CO_3^{2-}$	$10^{-3.2} = (Ca^{2+})(CO_3^{2-})/(CaCO_3^0)$
$NaHCO_3^0 = Na^+ + HCO_3^-$	$10^{+0.25} = (Na^+)(HCO_3^-)/(NaHCO_3^0)$
$CaSO_4^0 = Ca^{2+} + SO_4^{2-}$	$10^{-2.31} = (Ca^{2+})(SO_4^{2-})/(CaSO_4^0)$
$KSO_4^- = K^+ + SO_4^{2-}$	$10^{-0.94} = (K^+)(SO_4^{2-})/(KSO_4^-)$
$NaSO_4^- = Na^+ + SO_4^{2-}$	$10^{-0.72} = (Na^+)(SO_4^{2-})/(NaSO_4^-)$
$MgHCO_3^+ = Mg^{2+} + HCO_3^-$	$10^{-1.18} = (Mg^{2+})(HCO_3^-)/(MgHCO_3^+)$
$MgCO_3^0 = Mg^{2+} + CO_3^{2-}$	$10^{-3.4} = (Mg^{2+})(CO_3^{2-})/(MgCO_3^9)$
$MgSO_4^0 = Mg^{2+} + SO_4^{2-}$	$10^{-2.36} = (Mg^{2+})(SO_4^{-2})/(MgSO_4^0)$

consistent with the chemical composition of seawater. However, the calculated results on Ca concentration, Mg concentration, carbonate alkalinity, P_{CO_2} and F concentration are considerably different from those of seawater. Kramer (1965) considered that this inconsistency is caused by large uncertainty of thermochemical data. However, this was not regarded as no attainment of chemical equilibrium. In addition to the problem of attainment of chemical equilibrium, the inconsistency between Kramer's calculations and seawater chemistry may be caused by the effect of thermochemical properties of solid solutions. For example, equilibrium constants for the reactions, (4.24) and (4.25)

4.2 Seawater System

$$\text{H-montmorillonite(C site)} + \text{Na}^+ = \text{Na-montmorillonite(C site)} + \text{H}^+ \quad (4.24)$$

and

$$\text{SrCO}_3 = \text{Sr}^{2+} + \text{CO}_3^{2-} \quad (4.25)$$

are written as

$$K_{4-24} = (a_{\text{Na-mont}} a_{\text{H}^+})/(a_{\text{H-mont}} a_{\text{Na}^+}) \quad (4.26)$$

and

$$K_{4-26} = (a_{\text{Sr}^{2+}} a_{\text{CO}_3^{2-}})/(a_{\text{SrCO}_3}) \quad (4.27)$$

where $a_{\text{Na-mont}}$, $a_{\text{H-mont}}$ and a_{SrCO_3} are activity of Na-montmorillonite component and H-montmorillonite component, SrCO$_3$ component in strontianite, respectively.

We have to use these relations for the calculations which are expressed as functions of activities, but previous workers have used the apparent equilibrium constants as functions of mole fraction and molality.

4.2.3 Ion Exchange Equilibrium

Abundant clay minerals are contained in marine sediments. Ion exchange reactions between clay minerals and seawater are important factor controlling chemical compositions of seawater. The generalized ion exchange reaction is expressed as

$$m_{As} + B_{aq} = m_{Bs} + A_{aq} \quad (4.28)$$

where s is solid phase, and aq is aqueous species.

Equilibrium constant for Eq. (4.28) (K_{4-28}) is

$$K_{4-28} = (a_{Aaq} a_{Bs})/(a_{As} a_{Baq}) \quad (4.29)$$

Ion exchange reaction between clay minerals and seawater is generally very fast, resulting to ion exchange equilibrium. Therefore, chemical composition of seawater is greatly affected by ion exchange equilibrium (Kramer 1965; Stumm and Morgan 1970; Sayles and Margelsdorf 1977). Ion exchange equilibrium constants are presented in Table 4.3.

Table 4.3 Equilibrium constants of cation exchange reactions between clay minerals and aqueous solution (Lerman 1979)

Cation exchange reaction	Mineral (X)	Equilibrium constant log K
$Na_2X_2 \to H_2X_2$	Montmorillonite	6.452
$Na_2X_2 \to Li_2X_2$	Montmorillonite	−0.059
	Bentonite	−0.035
	Vermiculite	−1.059
$Na_2X_2 \to K_2X_2$	Bentonite	0.449
	Beidellite	1.598
$Na_2X_2 \to Rb_2X_2$	Bentonite	0.930
	Bentonite	3.400
$Na_2X_2 \to Cs_2X_2$	Montmorillonite	3.16–3.34
	Bentonite	1.585
$Na_2X_2 \to MgX_2$	Vermiculite	−0.194
$Na_2X_2 \to CaX_2$	Vermiculite	0.009
$Na_2X_2 \to Sr$	Montmorillonite	0.243
	Vermiculite	−0.009
$Na_2X_2 \to Ba$	Montmorillonite	0.040
$KX_2 \to Li_2X_2$	Montmorillonite	−1.796
$CaX_2 \to K_2X_2$	Montmorillonite	1.06–1.82
	Vermiculite	1.51–1.58
$CaX_2 \to SrX_2$	Bentonite	0.113
	Vermiculite	0.119
$BaX_2 \to MgX_2$	Vermiculite	−0.271
$BaX_2 \to CaX_2$	Vermiculite	−0.053
$BaX_2 \to SrX_2$	Vermiculite	−0.006

4.2.4 Factors Controlling Chemical Composition of Seawater (Input and Output Fluxes)

Chemical composition of seawater is controlled by various processes. River water input and sedimentation have been considered as important processes controlling chemical composition of seawater. However, there are other processes controlling chemical composition of seawater. They are seawater cycling at mid-oceanic ridge, input of volcanic gas, evaporation, weathering of oceanic crust, aerosol fall, anthropogenic pollution etc. (Wolery and Sleep 1976; Holland 1978).

4.2.4.1 River Water

Input of river water to seawater is the most important process controlling chemical composition of seawater. Chemical composition of river water varies widely (Table 4.4). Thus, it is difficult to estimate average chemical composition of river water in the world. However, it can be regarded as average chemical composition of major river water in continent (Missippi, Naile, Yangtze river etc.). Chemical composition of major river water is determined by water–rock interaction, mixing

4.2 Seawater System

Table 4.4 Chemical composition of river water (mg L^{-1}) (Nishimura 1991)

Component	Japanese 225 average riverwater Kobayashi (1960)	Japanese 42 average riverwater Sugawara (1967)	Average world Livingstone (1963)	Average world Martin et al. (1979)	As suspended solid Martin et al. (1979)
Evaporation residue	74.8				
Suspended solid	29.2				400
Cl$^-$	5.8	5.2	7.9		
HCO$_3^-$	31.0		58.4		
SO$_4^{2-}$	10.6	10	11.2		
Na$^+$	6.7	5.1	6.3	5.1	2.8
K$^+$	1.19	1.0	2.3	1.35	8.0
Ca^{2+}	8.8	6.3	15	14.6	8.6
Mg^{2+}	1.9	2.4	4.1	3.8	4.7
Fe	0.26		0.2		
N	0.26		0.2		
Ammonia N	0.05				
P	0.02				
Dissolved SiO$_2$	19.0	17	13.1	11.6	244

ratio of ground water to surface water, evaporation, biological activity, and aerosol fall. River water is a mixture of ground water and surface water (rainwater etc.). If chemical composition of river water is influenced greatly by deep ground water, chemical composition of river water is close to the equilibrium with rocks. Figure 4.12 shows the relationship between HCO$_3^-$ concentration and Ca^{2+} concentration which are in equilibrium with calcite as a function of P$_{CO_2}$. It is found in Fig. 4.12 that chemical composition of average world river waterplot close to the point representing equilibrium between atmospheric CO$_2$(P$_{CO_2}$ = 10$^{-3.5}$ atm) and calcite. Chemical reactionbetween calcite and aqueous solution is very rapid compared with that between silicatesand aqueous solution. This implies that the chemical equilibrium between silicates and aqueous solution is difficult to be attained.

The chemical compositions of major river water (Colorado, Missippi, Yukon, Columbia, Ganges) are plotted in the chemical weathering dominated region on Na/(Na + Ca)—total dissolved concentration diagram (Gibbs 1970). Chemical weathering depends on rock types (Walling 1980; Meybeck 1986; Stallard and Edmond 1987). The chemical compositions of river water are different in silicates-dominated rocks (igneous rocks etc.), carbonates-dominated rocks (limestone, dolomite rock), and evaporite.

Recently, detailed studies on the factors controlling river water chemistry such as biological activity (Likens et al. 1987), soilwater (Holland 1978; Likens et al. 1987) and geography (Drever 1988) have been done. However, the chemical compositions of major river water plot in chemical weathering-dominated region.

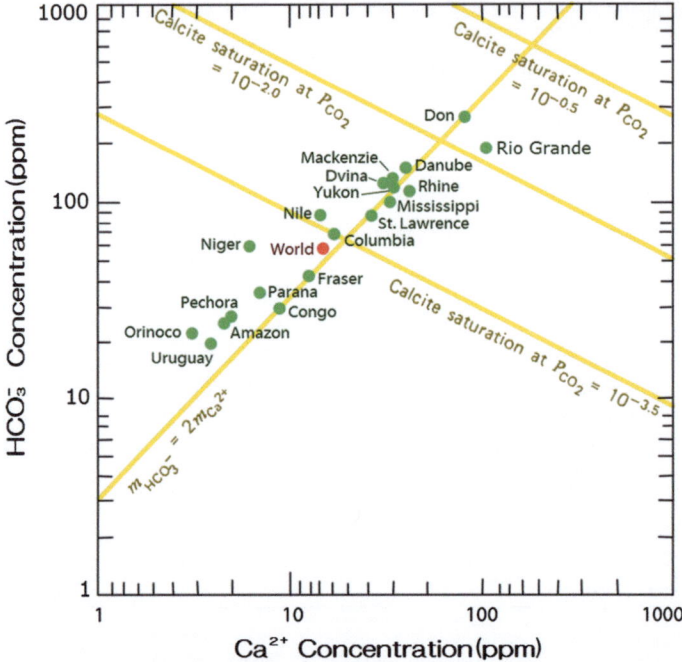

Fig. 4.12 The relationship between the bicarbonate and calcium concentrations in some major rivers (Livingstone 1963; Holland 1978)

Therefore, the relationship between chemical weathering and river water chemistry is considered below.

Gaillardet et al. (1999) estimated contribution of chemical weathering to river water chemistry based on Sr isotopic composition, concentration ratio of elements to Na, and runoff. According to their works, contributions of chemical weathering for Si and K are 100 % and 60 %, respectively. Those for Na, Ca, HCO_3^- and Sr are less than 50 %. It is obtained that the contribution of Si is the highest. Therefore, Si concentration of river water is considered below.

Chemical weathering relates to runoff, rainfall and temperature. The relationship between runoff and concentration of river water will be derived below.

Holland (1978) summarized analytical data on major river water in a world and showed that runoff relates to chemical composition. Shikazono (2002) derived this relationship based on dissolution kinetics-fluid flow model which is given below.

The change in concentration with time for perfectly mixing model is represented by

$$dm/dt = k(A/M)(m_{eq} - m)/m_{eq} + (q/V)(m_i - m) \quad (4.30)$$

where m is concentration, t is time, k is rate constant, A is surface area, M is mass of water, m_{eq} is equilibrium concentration, q is volume flow rate, v is volume of system as water, and m_i is input concentration.

4.2 Seawater System

Assuming steady state is attained and $m_i = 0$, we obtain

$$k(A/M)(m_{eq} - m)/m_{eq} - (q/V)m = 0 \qquad (4.31)$$

Runoff (Δf) is given by

$$\Delta f = q/A' \qquad (4.32)$$

where A' is watershed area which is approximated by

$$A' = V'/h \qquad (4.33)$$

where h is thickness of reservoir, and V' is volume of reservoir.
V' is related to V (volume of aqueous solution in reservoir) as

$$V' = V\o \qquad (4.34)$$

where ø is porosity.
Therefore

$$\Delta f = q/A' = qh/V' = q\o h/V \qquad (4.35)$$

Combination of this equation with Eq. (4.31) yields

$$k(A/M)(m_{eq} - m)/m_{eq} - (\Delta f/h)m = 0 \qquad (4.36)$$

Therefore, we obtain

$$m = m_{eq}/\{1 + \Delta f m_{eq}/[\Delta hk(A/M)]\} \qquad (4.37)$$

If $\Delta f m_{eq}/\{\Delta hk(A/M)\}$ is negligibly small compared with 1

$$m = m_{eq} \text{ and } \log m = \log m_{eq} \qquad (4.38)$$

In this case concentration is equal to equilibrium concentration.
If $\Delta f m_{eq}/\{\Delta hk(A/M)\}$ is very high compared with 1

$$m = \Delta hk(A/M)/\Delta f \qquad (4.39)$$

and

$$\log m = -\log \Delta f + \log\{\Delta hk(A/M)\} \qquad (4.40)$$

This equation means concentration decreases with increasing Δf at constant Δhk (A/M). The relationship between Si concentration and runoff will be derived below.

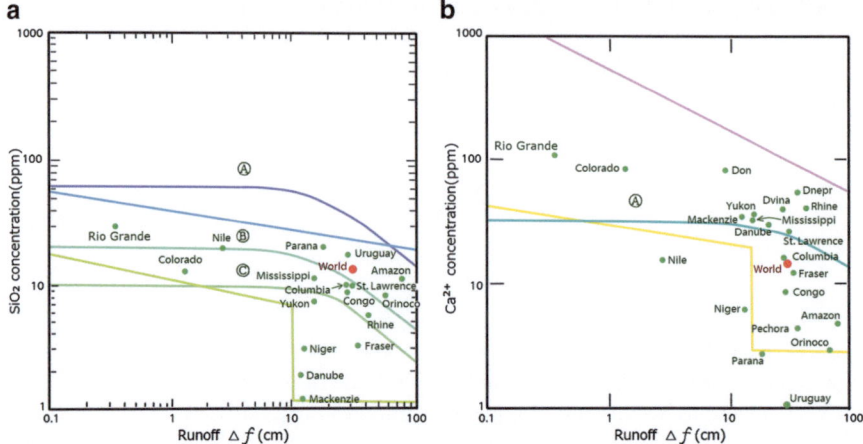

Fig. 4.13 (**a**) Relationship between H_4SiO_4 concentration and runoff obtained based on a rock–water reaction kinetics-fluid flow model. *A*: Solubility of Ca·Na-feldspar (low Δf region), $\emptyset hk(A/M) = 10^{-12.5}$ (high Δf region), $P_{CO_2} = 10^{-3.5}$ atm, *B*: solubility of chalcedony(low Δf region), $\emptyset hk(A/M) = 10^{-12.5}$ (high Δf region), *C*: solubility of quartz(low Δf region), $\emptyset hk (A/M) = 10^{-12.5}$ (high Δf region). (**b**) Relationship between Ca^{2+} concentration and runoff obtained on the basis of a rock–water reaction kinetics-fluid flow model. *A*: Solubility of Ca·Na-feldspar (low Δf region), $\emptyset hk(A/M) = 10^{-11.5}$ (high Δf region), $P_{CO_2} = 10^{-3.5}$ atm

Si concentration of river water is controlled by dissolution of Ca·Na-feldspar, which is represented by

$$Na_{0.5}Ca_{0.5}Al_{1.5}Si_{2.5}O_8(\text{feldspar}) + 1.5H^+ + 11/2H_2O$$
$$\rightarrow 0.5Na^+ + 0.5Ca^{2+} + 3/4Al_2Si_2O_5(OH)_4(\text{kaolinite}) + H_4SiO_4 \quad (4.41)$$

The relationship between H_4SiO_4 concentration determined by this reaction and runoff is derived using Eq. (4.40) (Fig. 4.13).

Rate constant, k, for Na·Ca-feldspar is $10^{-11.5}$ (Sverdrup 1990). The calculated relationships between H_4SiO_4 concentration and runoff shown in Fig. 4.13 are roughly consistent with analytical data. In small runoff region, Si concentration is relatively constant, but this decreases with increasing runoff in high runoff region. If we compare the theoretical curve for Ca^{2+} concentration with analytical data, it is found that both trends are roughly consistent with each other. However, Ca^{2+} concentration varies widely in small runoff region. This variation may be related to high dissolution rate of carbonates.

Holland (1978) summarized the data on river water from small runoff region and these data plot on HCO_3^-–Ca^{2+} concentration diagram.

Average world river water composition plots close to the cross point of calcite saturation line for $P_{CO_2} = 10^{-3.5}$ atm and electroneutrality line which is approximated as $m_{HCO_3^-} = 2m_{Ca^{2+}}$. This shows that river water with small runoff is in equilibrium with calcite. Therefore, it is inferred that chemical composition of river

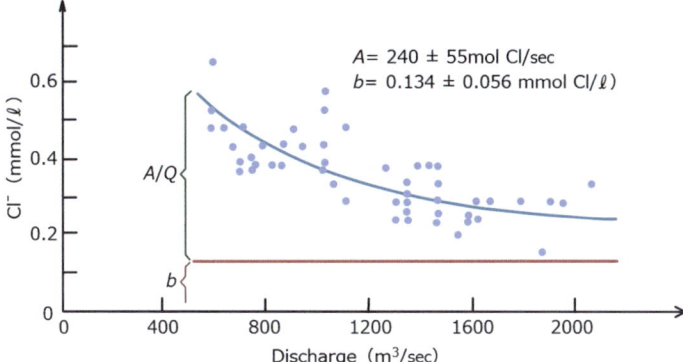

Fig. 4.14 The relationship between Cl concentration and discharge (Drever 1988)

water in different geologic environment (silicate-dominated and carbonate-dominated region) is distinct. Walling (1980) pointed out that rock type controls chemical composition of river water. Maximum dissolved solids concentration in river water at 700 mmL year^{-1} annual runoff decreases in the order of limestone → igneous rocks → sandstone, schist, gneiss → granite. This order depends on high dissolution rate for carbonates and low for silicates.

As noted above, Si and Ca concentrations of river water are highly dependent on rock–water interaction. However, some elements are not dependent on rock–water interaction. For example, Cl$^-$ concentration inversely correlates with runoff (Fig. 4.14) (Drever 1988), indicating that aerosol fall from atmosphere controls the concentrations of Cl$^-$ in river water. However, dissolution of evaporate (NaCl, KCl) controls Cl$^-$ concentration in the region where evaporate occurs widely.

Lasaga et al. (1994) derived the relationship between Si concentration and runoff based on the following relation.

$$dm/dt = k(A/M) - v(dm/dX) = 0 \qquad (4.42)$$

where m is Si concentration, A/M is specific surface area/mass of water ratio, k is dissolution rate constant, v is flow rate of river, and X_{reg} is average distance of water migration in regolith.

From the following equation

$$(A/M)kX_{reg}/v = m \qquad (4.43)$$

and taking logarithm we obtain

$$\log m = -\log v + \log\{(A/M)kX_{reg}\} \qquad (4.44)$$

Figure 4.15 shows the relationship between Si concentration and Darcy's velocity (Lasaga et al. 1994). These theoretical curves can roughly explain analytical

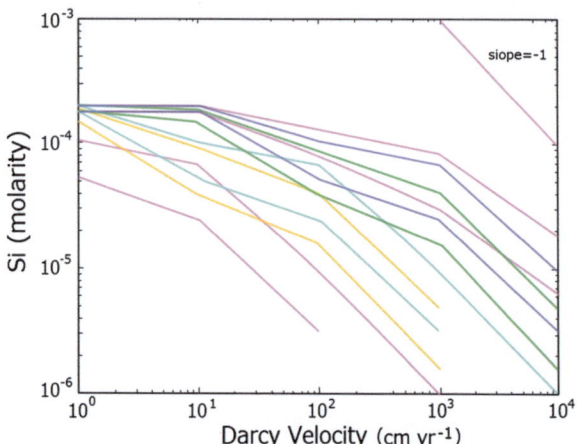

Fig. 4.15 Relationship between Darcy velocity (runoff) and Si concentration (Lasaga et al. 1994). 20 % Na-feldspar, porosity = 0.1, initial pH = 4.5

data on Si in river water in basalt region (e.g., Hawaii, Columbia basalt plateau, Iceland).

Cation and anion concentrations of river water are in an order, $Ca^{2+} > Mg^{2+} > Na^+$ and $HCO_3^- > SO_4^{2-} > Cl^-$. Seawater contains cations and anions as $Na^+ > Mg^{2+} > Ca^{2+}$ and $Cl^- > SO_4^{2-} > HCO_3^-$, respectively that are different from those of river water. When river water inputs to seawater, it mixes with seawater, resulting to a removal of base metals from mixed solution to sediments. For example, base metal elements in polluted river water inputting to Tokyo Bay, and organic matter settle onto seawater bottom in the bay. The removal of base metals and other elements from seawater near the coast significantly influences chemical composition of seawater.

4.2.4.2 Formation of Minerals

Fine particles, colloids and dissolved species are transported by river water to seawater. Fine particles dissolve in seawater and minerals precipitate and settle on to sea bottom. Mackenzie and Garrels (1966) calculated the fluxes transported by river water to ocean and removal fluxes by the formation of minerals from ocean to sea bottom (Table 4.5).

The mail minerals formed in seawater include halite (NaCl), Na-montmorillonite ($Na_{0.33}Al_{2.33}Si_{3.67}O_{10}(OH)_2$), Mg-chlorite ($Mg_5Al_2Si_3O_{10}(OH)_2$), K-illite ($KAl_3Si_3O_{10}(OH)_2$), calcite ($CaCO_3$), Mg-calcite ($(Ca,Mg)CO_3$), anhydrite ($CaSO_4$), gypsum ($CaSO_4 \cdot 2H_2O$), amorphous silica ($SiO_2$) and barite ($BaSO_4$). Various processes lead to the formations of the minerals. For example, rock salt (NaCl), calcite, gypsum and Mg-sulfate form as evaporite. Clay minerals such as montmorillonite and illite form by submarine weathering. Calcite, aragonite, Mg-calcite and amorphous silica form accompanied by biogenic activities. Barite forms in marine sediments. Ba^{2+} dissolves from radiolaria, feldspar and volcanic

Table 4.5 Mackenzie and Garrels (1966) mass balance calculation for the removal of river-derived constituents from the ocean (Holland 1978)

Step. No.	Reaction (balanced in terms of mmol of constituents used)	Constituent balance ($\times 10^{21}$ mmol)								HCO_3^- Consumed (−) Evolved (+)	CO_2 Consumed (+) Evolved (−)	Products ($\times 10^{21}$ mmol)	Percentage of Total Products Formed (mol basis)
		SO_4^{2-}	Ca^{2+}	Cl^-	Na^+	Mg^{2+}	K^+	SiO_2	HCO_3^-				
	Amount of material to be removed from ocean in 10^3 year ($\times 10^{21}$ mmol)	382	1,220	715	900	554	189	710	3,118				
1	$95.5FeAl_6Si_6O_{20}(OH)_4 + 191SO_4^{2-} + 47.8CO_2 + 55.7C_6H_{12}O_6 + 238.8H_2O = 286.5Al_2Si_2O_5(OH)_4 + 95.5FeS_2 + 382HCO_3^-$	191	1,220	715	900	554	189	710	3,500	+382	−48	96 pyrite 287 kaolinite	3 8
2	$191Ca^{2+} + 191SO_4^{2-} = 191CaSO_4$	0	1,029	715	900	554	189	710	3,500			191Cz	
3	$52Mg^{2+} + 104HCO_3^- = 52MgCO_3 + 52CO_2 + 52H_2O$	0	1,029	715	900	502	189	710	3,396	−104	+52	52 magnesite in calcite	
4	$1029Ca^{2+} + 2058HCO_3^- = 1029CaCO_3 + 1029CO_2 + 1029H_2O$	0	0	715	900	502	189	710	1,338	−2,058	+1,029	1,029 Calcite and/or aragonite	
5	$715Na^+ + 715Cl^- = 715NaCl$	0	0	0	185	502	189	710	1,338			715 NaCl	20
6	$71H_4SiO_4 = 71SiO_{2(s)} + 142H_2O$	0	0	0	185	502	189	639	1,338			71 Free silica	2
7	$138Ca_{0.17}Al_{2.33}Si_{3.67}O_{10}(OH)_2 + 46Na^+ = 138Na_{0.33}Al_{2.33}Si_{3.67}O_{10}(OH)_2 + 23.5Ca^{2+}$	0	24	0	139	502	189	639	1,338			138 Na-montmorillonite	4
8	$24Ca^{2+} + 48HCO_3^- = 24CaCO_3 + 24CO_2 + 24H_2O$	0	0	0	139	502	189	639	1,200	−48	+24	24 calcite and/or aragonite	1
9	$486.5Al_2Si_2O_{5.8}(OH)_4 + 139Na^+ + 361.4SiO_2 + 139HCO_3^- = 417Na_{0.33}Al_{2.33}Si_{3.67}O_{10}(OH)_2 + 139CO_2 + 625.5H_2O$	0	0	0	0	502	189	278	1,151	−139	+139	138 Na-montmorillonite	12
10	$100.4Al_2Si_2O_{5.8}(OH)_4 + 502Mg^{2+} + 60.2SiO_2 + 1004HCO_3^- = 100.4Mg_5Al_2Si_3O_{10}(OH)_8 + 1004CO_2 + 301.2H_2O$	0	0	0	0	0	189	218	147	−1,004	+1,004	100 chlorite	3
11	$472.5Al_2Si_2O_{5.8}(OH)_4 + 189K^+ + 189SiO_2 + 189HCO_3^- = 378K_{0.5}Al_{2.5}Si_{3.5}O_{10}(OH)_2 + 189CO_2 + 661.5H_2O$	0	0	0	0	0	0	29	−42	−189	+189	378 illite	11

Table 4.6 Sequence of precipitation of salts by evaporation of seawater (Hardie 1991)

Flux ratio	Precipitation sequence
0.96	Calcite–gypsum–anhydrite–$Na_2Ca(SO_4)_2$–NaCl–$K_2Ca_2Mg(SO_4)_4 \cdot 2H_2O$–$MgSO_4 \cdot 7H_2O$–$MgSO_4 \cdot 6H_2O$–$MgSO_4 \cdot H_2O$–$MgSO_4 \cdot KCl \cdot 11/4H_2O$–$Mg_2Cl_2 \cdot KCl \cdot 6H_2O$–$MgCl_2 \cdot 6H_2O$
1.05	Calcite–gypsum–anhydrite–NaCl–$K_2Ca_2Mg(SO_4)_4 \cdot 2H_2O$–KCl–$MgSO_4 \cdot KCl \cdot 11/4H_2O$–$MgSO_4 \cdot H_2O$
1.25	Calcite–gypsum–anhydrite–NaCl–KCl–$Mg_2Cl_2 \cdot KCl \cdot 6H_2O$–$CaCl_2 \cdot 6H_2O$–$CaCl_2 \cdot 2MgCl_2 \cdot 12H_2O$

Flux ratio: flux ratio of ridge hydrothermal solution to river water

glass combines with SO_4^{2-} which is present in very shallow part in the sediments to form barite. Interstitial seawater at very shallow part of sediments is saturated with respect to barite but seawater is not (Hanor 1969). Monnin et al. (1999) calculated the barite saturation state (the ratio of the ionic product to the solubility product) in the major world oceans for the GEOSECS stations (Ostlund et al. 1987). The results indicate that the Atlantic Ocean is undersaturated everywhere, but equilibrium is reached in a number of places; intermediate waters of the Indian and Pacific Oceans and surface waters of the Southern Ocean.

4.2.4.3 Formation of Evaporite

With the proceeding of evaporation of seawater, salts precipitate from seawater to form evaporite (Holland 1978; Harvie et al. 1980; Harolit 1991). Evaporite is sediment composed of Na, Ca, Mg and K-carbonates, sulfates and chlorides precipitated by evaporation at 1 atm and less than 100 °C. Evaporite is classified into three types according to the different source of aqueous solution.

(1) Freshwater + seawater, Na-K-CO_2-Cl-SO_4 type, mineral; Na_2CO_3.
(2) Freshwater + seawater, Na-K-Mg-Cl-SO_4 type, mineral; $MgSO_4$, Na_2SO_4.
(3) Hydrothermal solution + Na-K-Mg-Ca-Cl brine type, mineral; KCl ± $CaCl_2$, no occurrence of Na_2SO_4, $MgSO_4$.

The sequence of salt precipitation during evaporation of seawater depends on chemical composition of seawater and solubility of minerals. The sequence and amount of precipitation can be calculated using thermochemical data (Table 4.6, Fig. 4.16) (Harvie et al. 1980; Eugster et al. 1980) (Hardie 1991) based on Pitzer's ion interaction formalism (Pitzer 1973; Pitzer and Mayorga 1974).

4.2.4.4 Biological Activity

Chemical composition of seawater is significantly influenced by biological activity. For example, SiO_2 and $CaCO_3$ are formed by biological activity. Settling marine snow which is mainly composed of SiO_2 from dead organisms adsorbs dissolved species.

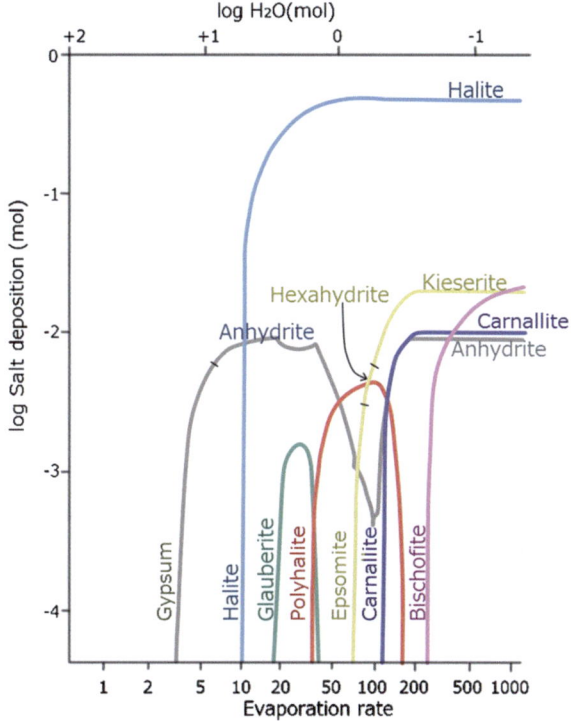

Fig. 4.16 Precipitated amount of minerals (salts) by evaporation of seawater (Hardie 1991)

SO_4^{2-} is reduced to H_2S by sulfate-reducing bacteria. Average concentration of Si in seawater is 29,000 µg g$^{-1}H_2O$, which is lower than that of river water (5,400 µg kg$^{-1}H_2O$). This decrease is due to biogenic formation of SiO_2. The concentration of Si in seawater is variable in a place to place (Wollast 1974). The concentration of Si in surface seawater is very low that is caused by biological activity. High concentration of Si in deep bottom seawater is due to the dissolution of SiO_2 shell of radiolaria and diatom.

Surface seawater is oversaturated with respect to $CaCO_3$ (calcite, aragonite) and biogenic $CaCO_3$ forms in surface seawater. Solubility of $CaCO_3$ increases with increasing pressure (depth). $CaCO_3$ dissolves below CCD (carbonate compensation depth).

SO_4^{2-} is reduced to H_2S using organic matter (e.g., lactic acid) and H_2. H_2S is fixed as iron sulfides (FeS, FeS_2).

Iron sulfide forms initially as precursor (amorphous FeS) that changes to pyrite during diagenesis. The mechanism of formation of pyrite is not well understood.

Framboidal pyrite is thought to be formed by the influence of bacteria. However, subhedral-euhedral pyrite forms by recrystallization of framboidal pyrite, precursor or direct formation by

$$Fe^{2+} + 2H_2S \rightarrow FeS_2 + 2H^+ + H_2 \qquad (4.45)$$

There are bacteria using energy liberated by the following reactions.

$$CO_2 + H_2S + O_2 + H_2O \rightarrow CH_2O + H_2SO_4 \tag{4.46}$$

$$2CO_2 + 6H_2 \rightarrow CH_2O + CH_4 + 3H_2O \tag{4.47}$$

Other bacteria such as sulfur-oxidizing bacteria, and hydrogen-oxidizing bacteria etc. (Jannash and Mottl 1985) are found near the vent of hydrothermal solution at mid-oceanic ridges and back arc basins. The biological activity by these bacteria causes the formation of Fe–Mn-minerals.

4.2.4.5 Interstitial Water

During the burial of seawater trapped in sediments, interaction of sediments and interstitial water, diffusion of dissolved species in interstitial water and advection of interstitial water occur, causing the variation in chemical composition of interstitial water.

Diffusion of dissolved species in interstitial water in sediments cause mass transport between seawater and interstitial water (Berner 1971, 1980; Lasaga and Holland 1976).

The concentration m (x, t) of an ion or compound produced during bacterial decay of organic matter in marine sediments not undergoing compaction is given by the equation.

$$\partial m(x,t)/\partial t = D_s \partial^2 m(x,t)/\partial x^2 - w \partial m(x,t)/\partial x + \partial m(x,t)/\partial t_{org} + \partial m(x,t)/\partial t \Big|_{org} \tag{4.48}$$

where m is the concentration of ion or dissolved species, x is the depth from sediment-seawater boundary, D_s is the diffusion coefficient of the ion or compound in the interstitial waters, w is sedimentation rate, and org is the organic matter.

If the decomposition of organic matter is a first-order process so that

$$dN(x,t)/dt = -kN(x,t) \tag{4.49}$$

where $N(x,t)$ is quantity of organic matter, then

$$dm(x,t)/dt_{org} = -\beta L k N(x,t) \tag{4.50}$$

where L is the mole ratio of the ion produced or consumed organic matter, oxidized and β is the conversion factor for the units of N to the units of m.

At steady state

$$0 = D_s \partial^2 m/\partial x^2 - w \partial m/\partial x + \beta L k N_0 exp\{-k(x/w)\} \tag{4.51}$$

Table 4.7 Elemental flux by the weathering of basalt (Wolery 1978)

Na^+	−1.3 to −7.7
K^+	−4 to −20
Ca^{2+}	10–50
Mg^{2+}	7–46
SiO^2	5–28
Mn	3–16
10 % of 200–800 m basalt is assured to be weathered	

For ions such as SO_4^{2-}, a boundary condition is $m(0,t) = m_0$. The solution of (4.51) is

$$m(x) = m_0 + (\beta L N_0 w^2)/(D_s k + \omega^2)\{1 - exp(-kw)x\} \quad (L < 0) \quad (4.52)$$

4.2.4.6 Low Temperature Seepage

Recently, low temperature seepages have been discovered at plate boundaries (Japan Trench, Nankai Trough, Oregon offshore, Peru offshore etc.). Low temperature seepages are reduced fluids with low concentration of SO_4^{2-}. Seepage water is originated from seawater and fresh (terrestrial) water. Seepage water is characterized by the presence of CH_4 produced by decomposition of organic matter and H_2S formed by reduction of seawater SO_4^{2-} and Cl^- concentration lower than seawater. Dehydration of hydrous minerals (mica, amphibole, clay minerals) may be also a source of seepage water (Tardy et al. 1991; Kastner et al. 1991; Ashi 1993).

Carbonate minerals precipitate by increasing pH due to the mixing of low temperature seepage and seawater. Carbonate may form chimney. Mn-bearing pyrite is found in a chimney, indicating a reducing condition of seepage (Shikazono 1994a, b). Kastner et al. (1991) considered the influence of seepage water on the global geochemical cycle.

4.2.4.7 Weathering of Oceanic Crust

Basalt below seafloor interacts with seawater at low temperatures (0–60 °C). Thickness of weathered basalt is several 100 m. Chemical and isotopic compositions of weathered basalt differ from those of fresh basalt. K_2O content and $\delta^{18}O$ of basalt increase with the proceeding of weathering. The increase in K_2O content is due to the formation of paragonite (yellowish gel-like material formed by the alteration of basalt glass).

Assuming that thickness of weathered basalt is 200–800 m and spreading rate of oceanic crust is 2.94 km^2 my^{-1} (Chase 1972), the amount of weathered basalt is estimated to be $(1.8–7.0) \times 10^{15}$ g $year^{-1}$ (Wolery 1978) (Table 4.7).

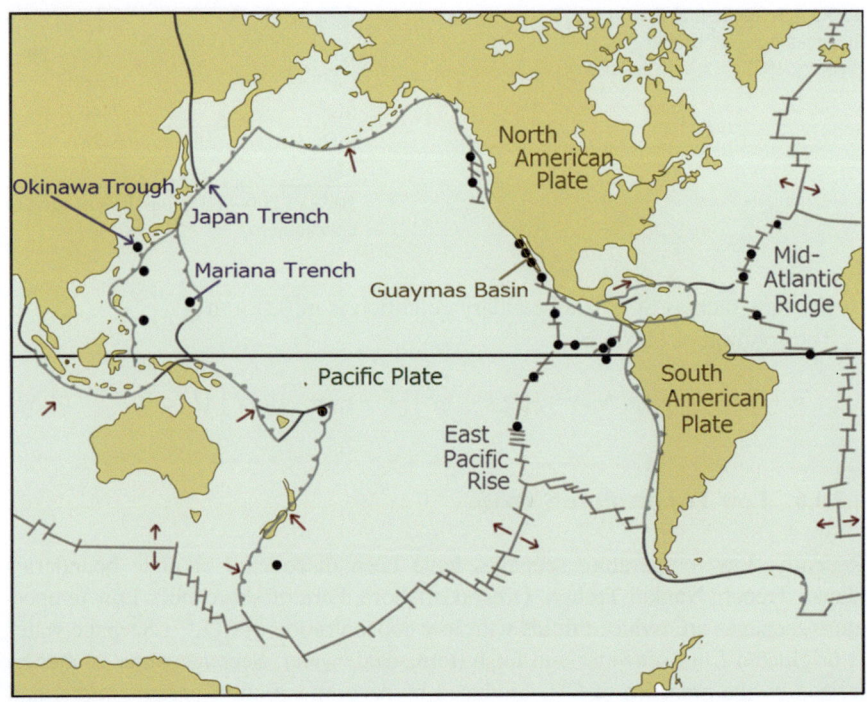

Fig. 4.17 Distribution of ore deposits at mid-oceanic ridge and back arc basin (Hannington et al. 1994)

Some elements are leached from basalt to interstitial water and seawater through basalt–seawater interaction. But some elements (Mn, Si, Ca) precipitate as veins in the weathered basalt and do not remove to seawater.

4.2.4.8 Hydrothermal Solution

Recently, many hydrothermal vents and formations of hydrothermal ore deposits have been discovered at mid-oceanic ridges and back arc basins (Fig. 4.17). Origin of hydrothermal solution is considered to be seawater from $\delta^{18}O$ and δD of hydrothermal solution. $\delta^{18}O$ and δD of hydrothermal solution are -0.1 to $+1.9$ ‰ and -0.9 to $+1.8$ ‰, respectively. They are close to those of seawater ($\delta^{18}O = 0$ ‰; $\delta D = 0$ ‰).

Slightly higher $\delta^{18}O$ values of hydrothermal solution than seawater value are caused by seawater–basalt interaction. Some elements (Mg, SO_4^{2-}) remove from circulating seawater to basalt by the formation of minerals (Mg-smectite, Mg-chlorite and anhydrite), but most of elements (base metals, Si etc.) are leached out from basalt.

4.2 Seawater System

Table 4.8 Comparison of primary axial high temperature hydrothermal chemical fluxes and riverine chemical fluxes (Elderfield and Schultz 1996)

Element	$C_{hydrotherma}$ fluid (mol/kg)[a]	$C_{seawater}$ (mol/kg)[a]	$F_{hydrothermal}$ (mol/year)	F_{rivers} (10^{10} mol/year)
Li	411–1,322 μ	26 μ	1.2–3.9 × 10^{10}	1.4
K	17–32.9 m	9.8 m	2.3–6.9 × 10^{10}	190
Rb	10–33 μ	1.9 μ	2.6–9.5 × 10^4	0.037
Cs	100–202 n	2.0 n	2.9–6.0 × 10^6	0.00048
Be	10–38.5 n	0	3.0–12 × 10^5	0.0037
Mg	0	53 m	−1.6 × 10^{12}	530
Ca	10.5–55 m	10.2 n	9.0–1,300 × 10^9	1,200
Sr	87 μ	87 μ	0	2.2
Ba	>8 to >42.6 μ	0.14 μ	>2.4 to 13 × 10^8	1.0
SO₄	0–0.6 m	28 m	−8.4 × 10^{11}	370
Alk	−0.1 to −1.0 m	2.3 m	−7.2 to 9.9 × 10^{10}	3,000
Si	14.3–22.0 m	0.05 m	4.3–6.6 × 10^{11}	640
P	0.5 μ	2 μ	−4.5 × 10^7	3.3
B	451–565 μ	416 μ	1.1–4.5 × 10^9	5.4
Al	4–20 μ	0.02 μ	1.2–6.0 × 10^8	6.0
Mn	360–1,140 μ	0	1.1–3.4 × 10^{10}	0.49
Fe	750–6,470 μ	0	2.3–19 × 10^{10}	2.3
Co	22–227 μ	0.03 n	6.6–68 × 10^5	0.011
Cu	9.7–44 μ	0.007 μ	3.0–13 × 10^8	0.50
Zn	40–106 μ	0.01 μ	1.2–3.2 × 10^9	1.4
Ag	26–38 n	0.02 n	7.8–11 × 10^5	0.0088
Pb	9–359 n	0.01 n	2.7–110 × 10^5	0.015
As	30–452 n	27 n	0.9–140 × 10^5	0.072
Se	1–72 n	2.5 n	3.0–220 × 10^4	0.0079
CO₂	5.7–16.7 m	2.3 m	1.0–12 × 10^{11}	
CH₄	25–100 μ	0 μ	0.67–2.4 × 10^{10}	
H₂	0.05–1 m	0 m	0.3–1.5 × 10^{10}	
H₂S	2.9–12.2 m	0 m	0.85–9.6 × 10^{11}	

[a] m = 10^{-3}; μ = 10^{-6}; n = 10^{-9}

Hydrothermal solution venting from seafloor influences seawater chemistry. Hydrothermal flux to seawater for each element can be estimated using chemical composition of hydrothermal solution and circulation rate of hydrothermal solution, and water flux by hydrothermal solution. Chemical composition of hydrothermal solution has been obtained by the analyses of hydrothermal solution, seawater–rock interaction experiments, and computer simulation on the changes in chemical compositions of hydrothermal solution interacted with rocks. The water flux and circulation rate of seawater and hydrothermal solution are estimated using heat flow at mid-oceanic ridges and ^{3}He/^{4}He of hydrothermal solution (Table 4.8) (Elderfield and Schultz 1996). However, it is difficult to estimate these average values because hydrothermal solution vents intermittently and not continuously. In early of 1980s when hydrothermal solution venting has been discovered, it has been thought that hydrothermal solution issuing from mid-oceanic ridges affects the seawater chemistry greatly (Edmond et al. 1979a). However, after that, Edmond et al.'s estimate was considered to be too large and hydrothermal flux from mid-oceanic ridges is about 1/6 of Edmond et al.'s estimate (Mottl 1983). Circulation of seawater and hydrothermal solution occurs not only at mid-oceanic ridges, but also at ridge flank (Mottl and Wheat 1994; Shikazono 1994a, b; Wheat and Mottl 2000). For example, Ca flux by hydrothermal solution at ridge flank is estimated to be $(0.15-0.30) \times 10^{19}$ mol my^{-1} that is larger than that from ridge axis $(0.15 \times 10^{17}$ mol my$^{-1})$ (Shikazono 1994a, b).

It has been thought that seawater chemistry is determined mainly by riverine flux and sedimentation flux before the discovering of hydrothermal venting. Most of elemental fluxes are balanced by these processes. However, Mg flux is unbalanced that is called "Mg problem" (Drever 1974). However, if we take into account Mg flux at mid-oceanic ridge axis and flank, the Mg problem could be solved. If seawater reacts with rocks at elevated temperature, Mg removes from seawater to rocks, resulting to very low concentration of Mg in hydrothermal solution, by the reaction

$$Mg^{2+}(\text{seawater}) + CaO(\text{rock})$$
$$\rightarrow MgO(\text{smectite, chlorite}) + Ca^{2+}(\text{modified seawater}) \quad (4.53)$$

SO_4^{2-} also removes from seawater to rocks as the precipitation of anhydrite due to the decreasing solubility of anhydrite with increasing temperature. The concentration of Ca^{2+} decreases due to the precipitation of anhydrite. However, it increases at more elevated temperatures by the reaction with basalt. Sr behaves similarly to Ca during seawater–rock interaction. K, Rb, Ba and base metal elements (Fe, Mn, Zn, Cu, Pb etc.) remove from rocks to hydrothermal solution at elevated temperatures due to increasing of solubility of minerals containing above elements. Hydrothermal solution vents and hydrothermal ore deposits form not only at mid-oceanic ridges but also at back arc basins (e.g., Okinawa Trough, Mariana Trough, Fiji) (Fig. 4.17). Present-day water flux by hydrothermal solution so far discovered from back arc basin is ca. 10–30 % of that from mid-oceanic ridges. However, at middle Miocene age, submarine hydrothermal activity at back arc basin such as Sea of Japan has been intense and Kuroko-type deposits formed associated with the hydrothermal activity (Shikazono 2003). Circulation of seawater and hydrothermal solutions at Japan Sea occurred, and caused intense and extensive hydrothermal alteration of submarine igneous rocks.

Figure 4.3 shows the negative correlation between CaO content and MgO content of altered igneous rocks in the Kuroko mining area. MgO content of altered rocks increases with the proceeding of seawater–rock interaction. MgO content of altered rocks represents extent of alteration of rock by seawater (which corresponds to water/rock ratio). Hydrothermal fluxes for the other elements due to seawater–rock interaction can be estimated from the relationship between MgO content and the eruption rate of igneous activity (Shikazono 1994a, b). Fine-grained Fe- and

Mn-hydroxides in hydrothermal plume mixed with seawater settle onto seafloor near and far away from the mid-oceanic ridge axis. Settling fine-grained Fe- and Mn-hydroxides adsorb some elements such as REE (rare earth elements) and P. This means that REE and P concentrations of seawater are affected by submarine hydrothermal activity.

Cited Literature

Ashi J (1993) Monthly Earth 15:636–640 (in Japanese)
Berndt ME, Seyfried WE Jr, Janecky DR (1989) Geochim Cosmochim Acta 53:2283–2300
Brimblecombe P, Hammer C, Rodhelt, Ryaboshapko A, Boutron CF (1989) In: Brimblecombe P, Lein AY (eds) Evolution of the global biogeochemical sulphur cycle. Wiley, New York, pp 77–121
Chase CG (1972) Geophys J Res Astron Soc 29:117122
Converse DR, Holland HD, Edmond JM (1984) Earth Planet Sci Lett 69:159–175
Drever JI (1974) In: Goldberg ED (ed) The sea, vol 5. Wiley, New York, pp 333–357
Edmond JM, Measure C, Magnum B, Grant B, Gordon LI, Corliss JB (1979a) Earth Planet Sci Lett 46:1–18
Edmond JM, Measure C, Mangum B, Grant B, Sclater FR, Collier R, Hudson A, Gordon LI, Corliss JB (1979b) Earth Planet Sci Lett 46:19–30
Elder J W (1966) N Z Dept Sci Industry Res Bull 169
Elderfield H, Schultz A (1996) Annu Rev Earth Planet Sci 24:191–224
Eugster HP, Harvie CE, Weare JH (1980) Geochim Cosmochim Acta 44:1335–1347
Feeley RA, Lewinson M, Massoth GJ, Rober-Baldo G, Lavelle JW, Byrne RH, Von Damm CV, Curl JRHC (1987) J Geophys Res 92:11347–11363
Gaillardet J, Dupre B, Louvat P, Allegre CJ (1999) Chem Geol 159:3–32
Gibbs RJ (1970) Science 170:1088–1090
Hanor JS (1969) Geochim Cosmochim Acta 33:894–898
Hardie LA (1991) Annu Rev Earth Planet Sci 19:131–168
Harolit LA (1991) Annu Rev Earth Planet Sci 19:131168
Harvie CE, Weare JH, Hardie LA, Eugster HP (1980) Science 208:498–500
Hayba DO, Bethke PM, Heald P, Foley NK (1985) In: Berger BR, Bethke PM (eds) Reviews in Economic Geology, vol 2, pp 129–168
Haymon RM, Kastner MC (1981) Earth Planet Sci Lett 53:363–381
Herzig PM, Becker KP, Stofeers P, Backer H, Bulum N (1988) Earth Planet Sci Lett 89:261–272
Jannash HW, Mottl MJ (1985) Science 229:717–725
Kastner M, Elderfield H, Martin JB (1991) Trans R Soc Lond A-335:261–273
Kawahata H (1989) Geochem J 23:255–268
Kawahata H, Shikazono N (1988) Can Min 26:555–565
Kramer JK (1965) Geochim Cosmochim Acta 29:921–945
Lasaga AC, Holland HD (1976) Geochim Cosmochim Acta 40:257–266
Lasaga AC, Soler JM, Ganor J, Burch T, Nagy KL (1994) Geochim Cosmochim Acta 58:23612386
Mackenzie FT, Garrels RM (1966) Am J Sci 264:507–525
Meybeck M (1986) Sci Geol Bull 39:3–77
Meybeck M (1987) Am J Sci 287:401–428
Monnin C, Hoareau G (2010) EMU Notes Min 10:227–258
Monnin C, Jeandel C, Cattaldo T, Dehairs F (1999) Mar Chem 65:253–261
Mottl MJ (1983) Geol Soc Am Bull 94:161180
Mottl MJ, Wheat CG (1994) Geochim Cosmochim Acta 58:2225–2237

Ogawa Y, Shikazono N, Ishiyama D, Sato H, Mizuta T, Nakano T (2007) Miner Deposita 42:219–233
Ohmoto H, Mizukami H, Drummond SE, Eldridge CS, Pisutha-Arnond V, Lenaugh TC (1983) Econ Geol Mon 5:570–604
Ostlund H, Craig H, Broecker D, Spencer D (1987) GEOSECS Atlantic, Pacific and Indians Ocean expeditions. Shorebased data and graphics. National Science Foundations, USA
Peter JM, Scott SD (1988) Can Min 26:567–587
Pitzer KS (1973) J Phys Chem 77:268–277
Pitzer KS, Mayorga G (1974) J Solution Chem 3:539–546
Sayles FC, Margelsdorf PC Jr (1977) Geochim Cosmochim Acta 41:951–960
Scott SD (1997) In: Barnes HL (ed) Geochemistry of hydrothermal ore deposits. New York, Wiley
Shikazono N (1988) Min Geol Spec Issue 12:47–55
Shikazono N (1994a) Geochim Cosmochim Acta 58:2203–2213
Shikazono N (1994b) Island Arc 3:59–65
Shikazono N (1998) J Geogr (Chigakuzasshi) 107:127–131; Hydrothermal (in Japanese with English abstract)
Shikazono N (2002) J Geogr 111:55–65 (in Japanese with English abstract)
Shikazono N, Fujimoto K (1996) Chikyukagaku (Geochem) 30:91–97 (in Japanese with English abstract)
Shikazono N, Holland HD (1983) Econ Geol Mon 5:329–344
Shikazono N, Yonekawa N, Karakizawa T (2002) Res Geol 52:211–222
Shikazono N, Kawabe H, Ogawa Y (2012) Res Geol 62:352–368
Sillen LG (1961) In: Sears M (ed) Oceanography. AAAS, Washington DC, pp 549–581
Sillen JK (1967a) Science 156:1189–1197
Sillen JK (1967b) In: Equilibrium concepts in natural water system. Advances in Chemistry Series, No. 67. American Chemical Society, Washington DC, pp 57–69
Stallard RF, Edmond JM (1987) J Geophys Res 92:8293–8302
Tively MK, Mcduff RE (1990) J Geophys Res 95:12617–12637
Walling DE (1980) In: Gower AM (ed) Water quality in catchment ecosystems. Wiley, New York, pp 1–48
Wells JT, Ghiorso MS (1991) Geochim Cosmochim Acta 55:2467–2481
Wheat CG, Mottl MM (2000) Geochim Cosmochim Acta 64:629–642
Wolery TJ (1978) Some chemical aspects of hydrothermal processes at Midoceanic ridges: a theoretical study. I. Basaltsea water reaction and chemical cycling between the oceanic crust and the oceans. II. Calculation of chemical equilibrium between aqueous solutions and minerals. Ph.D. thesis Northwestern University, p 262
Wolery TJ, Sleep NH (1976) J Geol 84:249–275
Wollast R (1974) In: Goldberg ED (ed) The sea. Wiley, New York, pp 359–392

Further Reading

Berner RA (1971) Principles of chemical sedimentology. McGraw-Hill, New York
Berner RA (1980) Early diagenesis: a theoretical approach. Princeton University Press, Princeton
Drever JI (1988) The geochemistry of natural waters, 2nd edn. Prentice Hall, Englewoods Cliff
Holland HD (1978) The chemistry of the atmosphere and oceans. Wiley, New York/Chichester/Bribane/Toronto
Holland HD (1984) The chemical evolution of the atmosphere and oceans. Princeton University Press, Princeton
Lasaga AC (1997) Kinetic theory in the earth sciences. Princeton University Press, Princeton
Nielsen AE (1964) Kinetics of precipitation. Pergamon, Oxford

Shikazono N (2003) Geochemical and tectonic evolution of arc-back arc hydrothermal systems: Implication for the origin of Kuroko and epithermal vein-type mineralization and the global geochemical cycle. Developments in Geochemistry 8, Elsevier, Amsterdam

Stumm W, Morgan JJ (1970) Aquatic chemistry. Wiley, New York

Sverdrup HU (1990) The kinetics of base cation release due to chemical weathering. Lund University Press, Sweden

Yamanaka T (1992) Introduction to biogeochemistry. Gakkaishuppan Center, Tokyo (in Japanese)

Chapter 5
Geochemical Cycle

Earth system is composed of subsystems such as atmosphere, hydrosphere, lithosphere (geosphere), biosphere and humans. Each subsystem can be divided into several parts. For example, geosphere includes crust (oceanic crust and continental crust), mantle (upper mantle and lower mantle) and core (outer core and inner core). Here, each part which constitutes subsystem is called reservoir. Each reservoir interacts chemically and physically with each other. Two aspects of interaction are heat and mass transfer.

5.1 General Equation

The mass balance equation for reservoir i as functions of fluxes is given by

$$dM_i/dt = \sum_j F_{ji} - \sum_i F_{ij} \pm \sum_k F_{ki} \tag{5.1}$$

where i and j are number of reservoir (i. . . . . j), M_i is mass of reservoir i, F_{ji} is flux from reservoir j to reservoir i (input), F_{ij} is flux from reservoir I to reservoir j (output) and F_{ki} is production or consumption in reservoir i.

Equation (5.1) is simplified one because Eq. (5.1) is represented as functions of many unknown parameters. For a simplification, number of reservoir is reduced and it is assumed that flux is proportional to mass or concentration. Thus

$$F_{ij} = k_{ij} M_i \tag{5.2}$$

where k_{ij} is related to residence time (τ) as $k_{ij} = 1/\tau$.

Assuming steady state condition with regard to the change in mass with time, we obtain

$$\sum_j F_{ji} - \sum_i F_{ij} \pm \sum_k F_{ki} = 0 \tag{5.3}$$

Using Eqs. (5.2) and (5.3) and k_{ij} values can be deduced. Time required to reach steady state depends on physical (hydrodynamic) state of system. It differs for perfectly mixing, non-mixing, and piston flow systems (Lerman 1979). Average concentration of perfectly mixing system is given by

$$m = m_i\{1 - exp(-1/\tau)\} \tag{5.4}$$

where m is average concentration of system, m_i is input concentration, and τ is residence time.

For non-mixing system

$$m = m_i t/\tau \tag{5.5}$$

Usually, flux does not linearly correlate to mass expressed as Eq. (5.2). In natural system input flux changes oscillatory with time as (Lasaga et al. 1981)

$$dM/dt = a + bsin(\omega t) \tag{5.6}$$

where M is mass of element in reservoir, and a, b, and ω are constant.

Seasonal variation in photosynthetic activity is expressed as (Lasaga et al. 1981)

$$dM/dt = a + bsin(\omega t) - kM \tag{5.7}$$

Equation (5.7) is solved as (Holland 1978)

$$M(t) = [(M_o - a/k + b\omega/k^2 + \omega^2)exp(-kt) + a/k \\ + (b/k^2 + \omega^2)\{bsin(\omega t) - \omega cos(\omega t)\} \tag{5.8}$$

where M_o is M at $t = 0$.

If $t = 1/k$, Eq. (5.8) is converted into

$$M(t) = a/k + (b/k_2) + \omega^2)\{ksin(\omega t) - \omega cos(\omega t)\} \\ = a/k + \{b/(k^2 + \omega^2)^{1/2} sin(\omega t - \delta)/2\}) \tag{5.9}$$

where $\delta = cos^{-1}\{k/(k^2 + \omega^2)^{1/2}(0 \leq \delta < \pi/2)$

Equation (5.9) indicates that M varies oscillatory with time.

The change of mass with time for two reservoir model is expressed as (Lasaga and Kirkpatrick 1981)

$$dM_1/dt = -k_{12}M_1 + k_{21}M_2 \tag{5.10}$$

$$dM_2/dt = k_{12}M_1 - k_{21}M_2 \tag{5.11}$$

From Eqs. (5.10) and (5.11), and combining them with $M_1 + M_2 = M$ where M is constant, we obtain

$$M_1(t) = k_{21}(M_1^0 + M_2^0)/(k_{12} + k_{21}) + \{(k_{21}M_1^0 - k_{21}M_2^0)/(k_{12} + k_{21})\}\exp\{-(k_{12} + k_{21})t\} + M_{10} \tag{5.12}$$

$$M_2(t) = k_{12}(M_1^0 + M_2^0)/(k_{12} + k_{21}) + \{k_{21}(M_1^0 - M_2^0)/(k_{12} + k_{21})\}\exp\{-(k_{12} + k_{21})t\} + M_{20} \tag{5.13}$$

where M^0 is initial M.

Residence time (τ) for reservoir 1 and reservoir 2 at steady state condition is

$$\tau_1 = M_{1\,steady\,state}/(dM_1/dt)_{input}$$
$$= M_{1\,steady\,state}/k_{12}M_{1\,steady\,state} = 1/k_{12} \tag{5.14}$$

$$\tau_2 = 1/k_{21} \tag{5.15}$$

Differential equations for 3, 4......... n reservoirs can be solved in a similar manner.

5.2 Carbon Cycle

It is essentially important to know chemical states and amount of carbon in reservoirs in order to consider carbon cycle. Dominant chemical states are CO_2 in atmosphere, organic matter and carbonates in biosphere, HCO_3^-, CO_3^{2-}, and H_2CO_3 in hydrosphere, and carbonates and carbon in geosphere (Table 5.1). The amount of carbon in geosphere is very large, compared with that in atmosphere, hydrosphere and biosphere (Table 5.2). There are two types of carbon cycle, short-term cycle (biogeochemical cycle) and long-term cycle (geochemical cycle). In biochemical cycle carbon transfers between biosphere and other reservoirs. Long-term cycle is carbon transfer between crust, hydrosphere and atmosphere. Time and space scales for the long-term cycle are larger and longer than the short-term cycle.

Table 5.1 Dominant chemical state of carbon in each reservoir

Atmosphere	CO_2
Biosphere	Organic carbon
Ocean	HCO_3^-
Geosphere	Carbonates, graphite

Table 5.2 Data pertaining to the carbon cycle (Holland 1979)

The carbon content of important reservoirs	carbon content (in units of 10^{15} g)
1. Atmosphere	690
2. Terrestrial biosphere	450
3. Dead terrestrial organic matter	700
4. Marine biosphere	7
5. Dead marine terrestrial matter	3,000
6. Dissolved in seawater	40,000
7. Recycled elemental carbon in the lithosphere	20,000,000
8. Recycled carbon in the lithosphere	70,000,000
9. Juvenile carbon	90,000,000

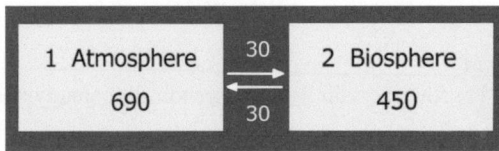

Fig. 5.1 Short-term carbon cycle between two reservoirs (atmosphere, biosphere) unit: 10^{15} g C, 10^{15} g C/y (Lasaga et al. 1981)

5.2.1 Short-Term Cycle (Biogeochemical Cycle)

Short-term cycle is represented by photosynthesis and decomposition of organic matter that is expressed as the reaction

$$CO_2 + H_2O \rightleftharpoons CH_2O + O_2 \tag{5.16}$$

From Fig. 5.1, k_{12} and k_{21} is obtained as $30/690 = 0.04348$ year^{-1} and $30/450 = 0.06667$ year^{-1}, respectively. Residence time is estimated to be $\tau_1 = 1/0.04348 = 23$ years and $\tau_2 = 1/0.06667 = 15$ years. The change of carbon mass with time in reservoir 1 and reservoir 2 in Fig. 5.1 is given by

$$\left. \begin{array}{l} M_1 = 690 + 50\exp(-0.11015t) \\ M_2 = 450 - 50\exp(-0.11015t) \end{array} \right\} \tag{5.17}$$

M_1 and M_2 in future can be predicted based on these equations. For example, carbon mass in atmosphere decreases to 707×10^{15} g, while carbon in biosphere increases to 433×10^{15} g and is same level after 20 years (696×10^{15} g in atmosphere, 444×10^{15} g in biosphere).

If 50 unit (one unit: 10^{15} g) carbon was added to atmosphere by burning of fossil fuel, and $M_{10} = 740$ and $M_{20} = 450$, carbon mass of reservoir represented as a function of time is

5.2 Carbon Cycle

$$\left.\begin{array}{l}M_1(t) = 720.3 + 19.7\exp(-0.11015t) \\ M_2(t) = 469.7 - 19.7\exp(-0.11015t)\end{array}\right\} \quad (5.18)$$

Steady state condition is attained, if $t \to \infty$ as $M_1 = 720.3$ and $M_2 = 469.7$. Carbon mass increases for both systems in steady state condition. In addition to reaction (5.16), the following reaction has to be taken into account for short-term carbon cycle.

$$Ca^{2+} + 2\,HCO_3^- \to CaCO_3 + CO_2 + H_2O \quad (5.19)$$

For example, both reactions, (5.16) and (5.19), occur in the formation of coral reef (Appendix, Plate 26) and carbonate shell of marine organisms (e.g. foraminifera, Appendix, Plates 27 and 28). In this case atmospheric CO_2 removes to coral reef by reaction (5.16), but also CO_2 releases from ocean to atmosphere by reaction (5.19). It depends on environmental condition (biogenic production rate, pH, dominant reaction etc.). CO_2 is generated by reaction (5.19). However, generated CO_2 changes to HCO_3^- according to reactions, $CO_2 + H_2O \to HCO_3^- + H^+$, $CO_2 + CO_3^{2-} + H_2O \to 2HCO_3^-$. If these reactions proceed considerably, CO_2 in atmosphere does not increase. Removal of atmospheric CO_2 occurs not only by coral reef, but also by other marine organisms (foraminifera etc.). Dead marine organisms settle down to deep sea. During the settling, the following reactions occur, resulting to uptake of atmospheric CO_2 and release of CO_2 to atmosphere.

$$CaCO_3 + CO_2 + H_2O \to Ca^{2+} + 2\,HCO_3^- \quad (5.20)$$

$$CH_2O + O_2 \to CO_2 + H_2O \quad (5.21)$$

Holland (1978) considered carbon geochemical cycle in five box model (Fig. 5.2). In this model, biosphere is divided into five boxes. The rate of photosynthesis of plants is related in P_{CO_2} by relation of the form

$$dM_{12}/dt = a\{1 - \exp(-bP_{CO_2})\} - c \quad (5.22)$$

where m is the respiration rate in the absence of CO_2 and a, b and c is constant. P_{CO_2} at the compensation point is

$$(P_{CO_2})_{comp} = -(1/b)\ln\{1 - (c/a)\} \quad (5.23)$$

The maximum rate of photosynthesis is

$$(dM_{12}/dt)_{max} = a - c \quad (5.24)$$

Fig. 5.2 The biological cycle of carbon; carbon contents are in units of 10^{15} g of C, and transfer rates in units of 10^{15} g of C/year^{-1} (Holland 1978). The data are largely from Bolin (1970), Reiners (1973), Koblentz-Mishke et al. (1970) and Whittaker and Likens (1973)

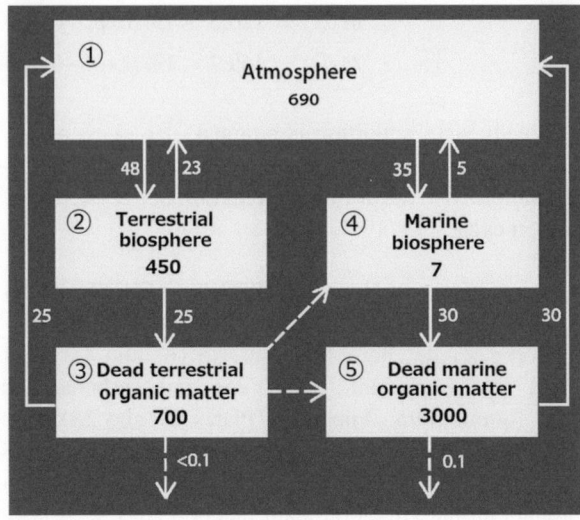

The rate of CO_2 return from the reservoir of dead organic matter to the atmosphere is assumed to be a linear function of the quantity of dead organic matter. Thus

$$dM_{12}/dt = eM_2 \tag{5.25}$$

$$dM_{23}/dt = fM_2 \tag{5.26}$$

$$dM_{31}/dt = gM_3 \tag{5.27}$$

where $e, f,$ and g are constant.

At steady state the following equations are established.

$$dM_{12}/dt - dM_{12}/dt - dM_{23}/dt = 0 \tag{5.28}$$

$$dM_{23}/dt - dM_{31}/dt = 0 \tag{5.29}$$

$$dM_{31}/dt - dM_{12}/dt + dM_{21}/dt = 0 \tag{5.30}$$

It follows that

$$a\left(1 - e^{-bP_{CO_2}}\right) - c - eM_2 - fM_2 = 0 \tag{5.31}$$

$$fM_2 - gM_3 = 0 \tag{5.32}$$

$$gM_3 - a\left(1 - e^{-bP_{CO_2}}\right) + c + eM_2 = 0 \tag{5.33}$$

5.2 Carbon Cycle

P_{CO_2} is proportional to amount of atmospheric CO_2 which is expressed as

$$bP_{CO_2} = b'M_1 \tag{5.34}$$

Therefore

$$M_2 = a\{1 - \exp(-b'M_1) - c\}/(e+f) \tag{5.35}$$

$$M_3 = (f/g)M_2 = (f/g)[a\{1 - \exp(-b'M_1) - c\}/(e+f)] \tag{5.36}$$

If the total amount of carbon ($M_1 + M_2 + M_3$) in the three reservoirs, is fixed, the

$$M_1 + M_2 + M_3 = M = constant \tag{5.37}$$

From Eqs. (5.35), (5.36) and (5.37), we obtain M_1, M_2 and M_3. From these values, fluxes between reservoirs and P_{CO_2} can be calculated.

5.2.2 Long-Term Cycle (Geochemical Cycle)

Long-term cycle occurs between geosphere (crust), atmosphere and hydrosphere. Long-term carbon transfer is controlled by dissolution and precipitation of carbonates and silicates. These reactions are given by

$$CaCO_3 + CO_2 + H_2O = Ca^{2+} + 2HCO_3^- \tag{5.38}$$

$$MgCO_3 + CO_2 + H_2O = Mg^{2+} + 2HCO_3^- \tag{5.39}$$

$$CaSiO_3 + 3H_2O + 2CO_2 = Ca^{2+} + H_4SiO_4 + 2HCO_3^- \tag{5.40}$$

$$MgSiO_3 + 3H_2O + 2CO_2 = Mg^{2+} + H_4SiO_4 + 2HCO_3^- \tag{5.41}$$

Reactions (5.38)–(5.41) indicate that CO_2 removes from the atmosphere to the hydrosphere by weathering of carbonate and silicate dissolution.

BLAG model (silicate-carbonate model) concerning long-term carbon cycle (Berner et al. 1983) is shown in Fig. 5.3. Basic reactions for BLAG model include reactions between CO_2 and Ca- and Mg-minerals (calcite, dolomite, Ca-silicate, Mg-silicate).

They are expressed as

$$CO_2 + CaSiO_3 \underset{\text{Igneous activity, Metamorphism}}{\overset{\text{Carbonate formation, Weathering}}{\rightleftharpoons}} CaCO_3 + SiO_2 \tag{5.42}$$

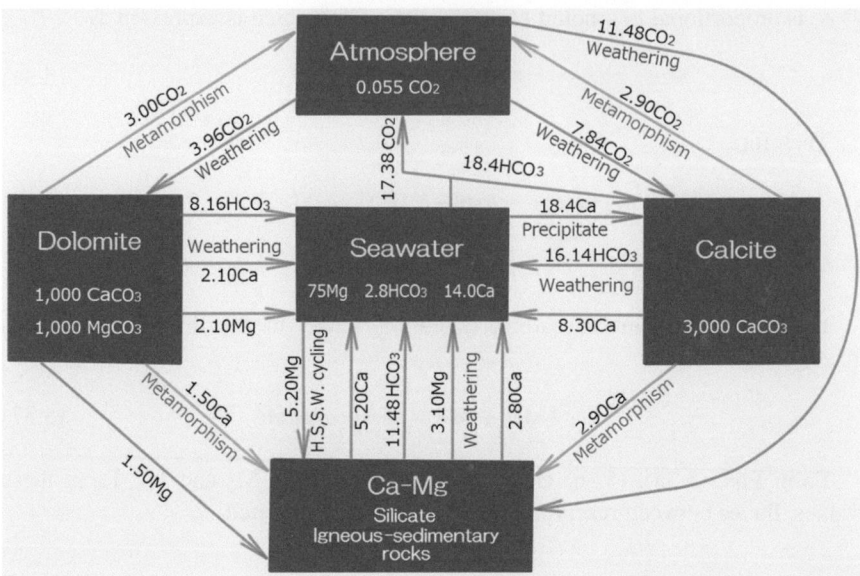

Fig. 5.3 Present-day carbonate-silicate cycle (Berner et al. 1983). Reservoir unit: 10^{18} mol, Flux unit: 10^{18} mol my^{-1}

In seawater following reaction occurs.

$$CO_2 + CaCO_3 + H_2O \underset{\text{Precipitation}}{\overset{\text{Dissolution}}{\rightleftharpoons}} Ca^{2+} + 2HCO_3^- \quad (5.43)$$

Generally magnesite ($MgCO_3$) and dolomite ($Ca, Mg\,(CO_3)_2$) do not precipitate from seawater due to their very slow precipitation rate. They form in the diagenetic process. The rate of Mg removal from seawater is by the precipitation of Mg-calcite ((Ca, Mg) CO_3). The Mg removal by the interaction of cycling seawater with oceanic basalt at mid-oceanic ridges relates to plate motion. Plate motion controls igneous and hydrothermal activity at subduction zone (Appendix, Plates 29, 30, 31, and 32). Weathering and riverwater flow (Appendix, Plate 33) rates depend on continental area, continental uplift rate and type of rocks. Seafloor spreading rate, production of dolomite and atmospheric temperature during the last one million years are known. Using these data and mass balance equations we can calculate the temporal variations in concentrations (atmospheric CO_2 concentration, Ca and Mg concentrations and pH of seawater), abundance of minerals and mass of carbon in each reservoir.

Giving the changes in seafloor spreading rate and continental area with time, amount of atmospheric CO_2 during the last one hundred million years compared with that of present-day atmospheric CO_2 was calculated by Berner et al. (1983) (Fig. 5.4) based on silicate-carbonate geochemical cycle model (BLAG model).

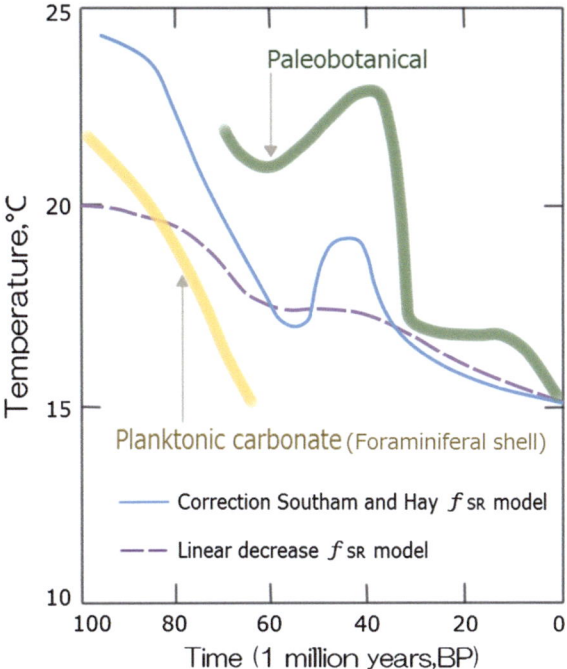

Fig. 5.4 Plots of worldwide mean annual atmospheric surface temperature versus time, based on BLAG model, compared with temperatures estimated from paleobotanical study and $\delta^{18}O$ of planktonic carbonate (35 °N) (Berner et al. 1983)

Figure 5.4 clearly indicates that atmospheric CO_2 concentration and temperature decreased during the last one hundred million years. This tendency is consistent with the estimate from paleobotanic study (Fig. 5.4).

Above results imply that long-term change of earth's surface environment (temperature, atmospheric CO_2 concentration) is greatly influenced by plate tectonics. The changes in plate tectonics cause the variations in seafloor spreading rate, cycling rate of seawater-hydrothermal solution at mid-oceanic ridges, continental area, and the rate of CO_2 degassing due to igneous, metasomatic and metamorphic activities at subduction zone.

BLAG model does not consider the cause for the tectonic change and assumes that CO_2 degassing rate at subduction zone linearly correlates to seafloor spreading rate. However, in fact, intensity of igneous activity at subduction zone during the last one hundred million years has intermittently changed and it does not correlate to ocean floor spreading rate (Kennet et al. 1977). Therefore, Kashiwagi and Shikazono (2003) (K&S model) (Fig. 5.5) proposed the global carbon cycle and climate model in which the rate of CO_2 degassing from subduction zone is not proportional to ocean floor spreading rate. K&S model considered the Cenozoic global CO_2 cycle including CO_2 flux due to hydrothermal and volcanic activities in the Miocene at a subduction zone (back arc basin and island arc) (Fig. 5.6). Their model results indicate that (1) the contribution of silicate weathering in the HTP (Himalaya and Tibetan Plateau) region is small; (2) the warming from late

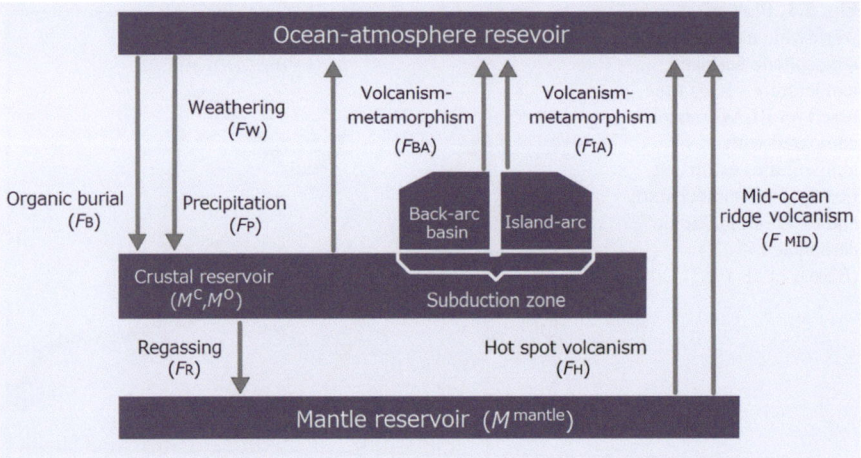

Fig. 5.5 Kashiwagi and Shikazono's model (K&S model) of the Cenozoic global CO_2 cycle (Kashiwagi and Shikazono 2003)

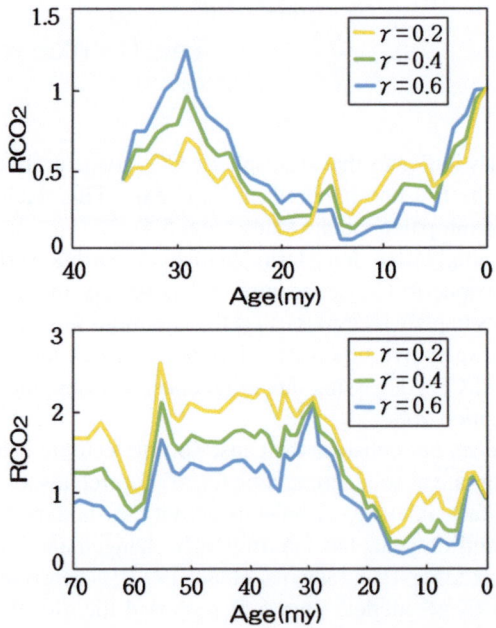

Fig. 5.6 (**a**) Atmospheric CO_2 variation estimated K&S model including volcanic eruption rate of Circum-Pacific region by Kennet et al. (1977) (Kashiwagi and Shikazono 2003). γ represents the contribution of the flux from back arc basin to that from subduction zones at present. $R_{CO_2} = P_{CO_2}^*/P_{CO_2}$, P_{CO_2}: present-day P_{CO_2}. (**b**) Atmospheric CO_2 variation estimated by modified BLAG model including CO_2 flux related to mantle plume activity (Kashiwagi et al. 2008). γ presents the contribution of the flux from back arc basin to that from subduction zones at present

5.3 Sulfur Cycle

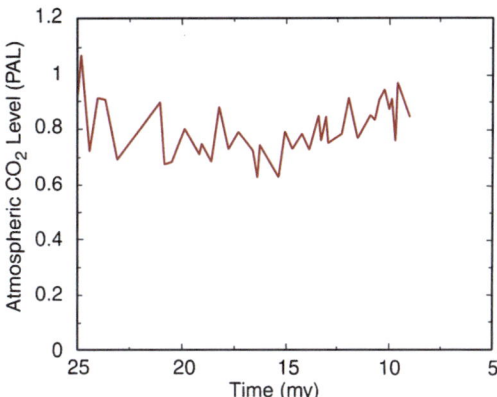

Fig. 5.7 P_{CO_2} estimates from Oligocene to late Miocene (modified after Pagani et al. 1999)

Oligocene to early Mioche might be due to the CO_2 from the back arc basin; (3) the cooling event in the middle Miocene (15 Ma) is caused by a large amount of the organic carbon burial; (4) their calculations agree with previous results based on $\delta^{13}C$ and δ B of foraminifera shell carbonates (Pagani et al. 1999; Pearson and Palmer 2000) (Figs. 5.7 and 5.8). This indicates that cooling occurred in the northern hemisphere, but the entire earth's surface temperature did not decline during the Miocene.

Kashiwagi and Shikazono (2003) modified GEOCARB-type model by Berner (1991, 1994) and Berner and Kothavala (2001) which is mathematically simpler but more complex geologically and biologically than BLAG model. By GEOCARB I model, the isotopic mass balance model of Garrels and Lerman (1984) was expanded to include the effect of changing area, elevation and position of the continent as well as evolution of land plants as they affect weathering rate and changes in sea floor area generation rate and the relative impotence of deep sea versus shallow platform carbonate deposition and atmospheric CO_2 level was calculated from a weathering feedback function for silicates which varies with time as vascular land plants arise and evolve.

5.3 Sulfur Cycle

Sulfur occurs abundantly in evaporite (5×10^{21} g), sediments (mainly shale, 2.7×10^{21} g), metamorphic rocks and igneous rocks (7×10^{21} g). Abundance of sulfur in seawater (1.3×10^{21} g) and atmosphere (3.6×10^{12} g) is small, compared with that in the rocks. Sulfur is present mainly as sulfate (anhydrite, $CaSO_4$, gypsum, $CaSO_4 \cdot 2H_2O$ etc.) in evaporite, pyrite (FeS_2), and gypsum ($CaSO_4 \cdot 2H_2O$) in sediments, sulfides (e.g., pyrite) and sulfates (barite, $BaSO_4$, anhydrite) in ore deposits, sulfate ion (SO_4^{2-}, $NaSO_4^-$, KSO_4^-) in seawater and SO_2 and H_2S in atmosphere.

Fig. 5.8 Atmospheric CO_2 data for the past 60 my (**a**) The entire record, (**b**) an enlargement of the past 25 my (Pearson and Palmer 2000)

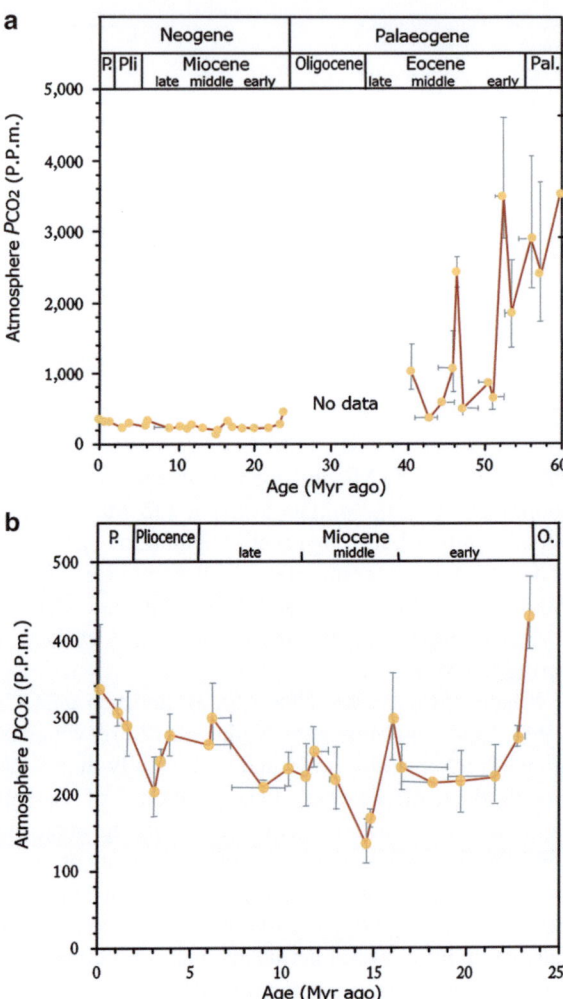

5.3.1 Short-Term Cycle

S in the atmosphere which is present as SO_2, H_2S, and organic sulfur (DMS, $(CH_3)_2S$, CS_2, COS etc.) oxidizes easily and removes by rainfall, implying that residence time of sulfur in the atmosphere is short. Activities of microorganism takes important role for short-term sulfur cycle. For example, DMS that is produced by plankton transfers from ocean to atmosphere (Lovelock et al. 1972). This sulfur flux is estimated as $(1.9–0.6) \times 10^{12}$ mol year^{-1} ($4 \pm 2 \times 10^{13}$ g year^{-1}). Activities of bacteria promote oxidation-reduction reactions in soils and affect the rate of sulfur transfer between soils and soil water. For example, sulfate-reducing bacteria and sulfur-oxidizing bacteria promote the following reactions.

5.3 Sulfur Cycle

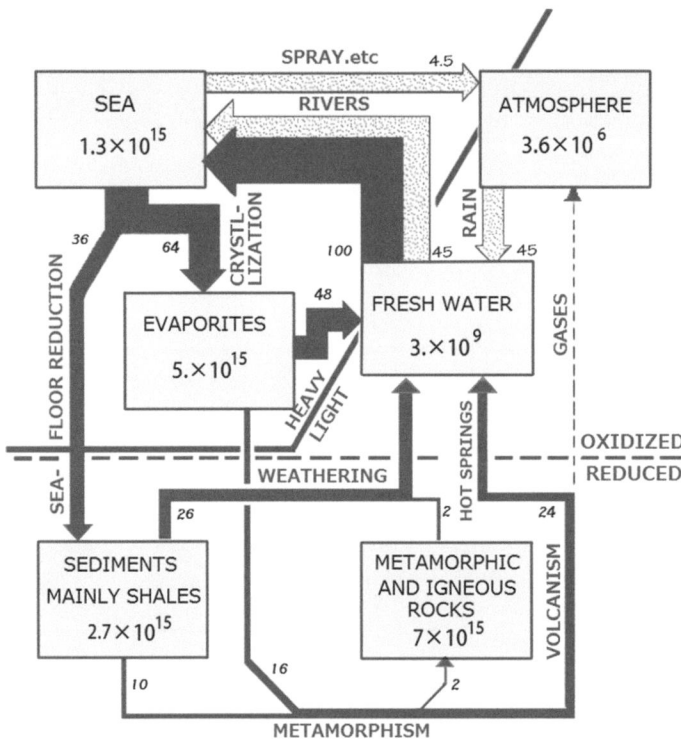

Fig. 5.9 Holser and Kaplan's (1966) representation of the geochemical cycle of sulfur. Masses are in metric tons. Most material above the *dashed line* is oxidized to sulfate, most below this line is reduced to sulfide; above the *solid line* heavy sulfur predominates ($\delta > +5$ ‰). Long-term (dark) and short-term fluxes of sulfur between reservoirs are indicated on the basis of 100 for the long-term component of fresh water sulfate flowing to the sea (Holland 1984)

$$SO_4^{2-} + 2H^+ \rightarrow H_2S + 2O_2 \tag{5.44}$$

$$H_2S + 2O_2 \rightarrow SO_4^{2-} + 2H^+ \tag{5.45}$$

If reaction (5.45) proceeds, pH decreases, resulting to release of elements from rocks and soils.

5.3.2 Long-Term Cycle

Long-term sulfur cycle between reservoirs is shown in Fig. 5.9. Important processes for long-term sulfur cycle include formation of sulfur-containing compounds (e.g., pyrite in sedimentary rocks, sulfates in evaporite), oxidation of sulfides in terrestrial environment, subduction of plate, emissions of volcanic gas and hydrothermal

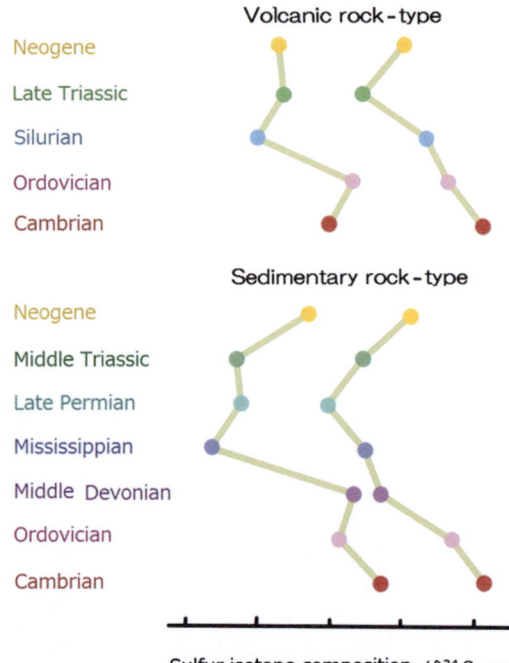

Fig. 5.10 Parallelism of sulfur isotopic compositions of stratabound sulfide deposits and seawater sulfate. *Solid circle* and *open circle* represents average sulfur isotopic composition of sulfide sulfur and sulfur isotopic composition of seawater sulfate sulfur estimated from evaporate (Sasaki 1979)

solution from crust and mantle and hydrothermal alteration of oceanic crust. Sulfur isotopic fractionation and the change in amount of sulfur in reservoirs occur, associated with these processes.

δ^{34}S of seawater and ore deposits (volcanic-hosted and sediment-hosted sulfide deposits) have been changed during the past geologic ages depending on the changes in sulfur fluxes between reservoirs (Fig. 5.10). The change in δ^{34}S of seawater can be estimated from δ^{34}S of evaporite. δ^{34}S of seawater is controlled by formation of pyrite from seawater, formation of evaporite, hydrothermal venting and input of river water to ocean. For example, light sulfur is taken up by biotic pyrite, resulting to higher δ^{34}S value of seawater, while input of hydrothermal solution with low δ^{34}S (+several ‰) reduces δ^{34}S value of seawater.

5.4 Coupled Geochemical Cycle: Sulfur–Carbon–Oxygen (S–C–O) Cycle

Geochemical cycle of each element interrelates. During recent years, large number of coupling geochemical modeling have been carried out (Kashiwagi et al. 2008 etc.).

5.4 Coupled Geochemical Cycle: Sulfur–Carbon–Oxygen (S–C–O) Cycle

Here classical work of Garrels and Lerman (1984) which treated sulfur–carbon–oxygen (S–C–O) global cycle is briefly shown below. The relationships between S, C, and O cycles are derived from the following reactions.

$$CH_2O + O_2 = CO_2 + H_2O \tag{5.46}$$

$$H_2S + O_2 = SO_2 + H_2O \tag{5.47}$$

Not only the chemical reactions involving gaseous and liquid phases such as (5.46) and (5.47) operating in S–C–O cycle, but also the reactions involving solid phases are important to control atmospheric CO_2 concentration. For example, the following reactions take important role as the factors controlling the atmospheric CO_2 concentration.

$$2FeS_2 + 3/2O_2 = Fe_2O_3 + 2S_2 \tag{5.48}$$

$$CaCO_3 + SO_2 + 1/2O_2 = CaSO_4 + CO_2 \tag{5.49}$$

$$CaMg(CO_3)_2 + SiO_2 = MgSiO_3 + CaCO_3 + CO_2 \tag{5.50}$$

Combination of above reactions leads to the following reaction (Garrels and Perry 1974)

$$\begin{aligned} 4FeS_2 + CaCO_3 + 7CaMg(CO_3)_2 + 7SiO_2 + 15H_2O \\ = 15CH_2O + 8CaSO_4 + 2Fe_2O_3 + 7MgSiO_3 \end{aligned} \tag{5.51}$$

$\delta^{34}S$ and $\delta^{13}C$ can be used to constrain the simulation on long-term S–C–O cycle. Amounts of C and S, $\delta^{34}S$ and $\delta^{13}C$ of reservoirs and fluxes between reservoirs today are shown in Fig. 5.11. Garrels and Lerman (1984) obtained evolutionary curves of $\delta^{34}S$ and $\delta^{13}C$ of Paleozoic seawater under the following assumptions.

(1) The amounts of S and C in atmosphere–ocean reservoir are constant. In Fig. 5.11 this is expressed as

$$F_{13} + F_{23} = F_{32} + F_{31}, F_{64} + F_{54} = F_{46} + F_{45} \tag{5.52}$$

The conditions of constant total mass can be written as

$$S_1 + S_2 + S_3 = S_t \tag{5.53}$$

(2) Weathering flux to the ocean (F_{13}, F_{23}, F_{54}, F_{64}) correlates linearly with mass in the reservoir.

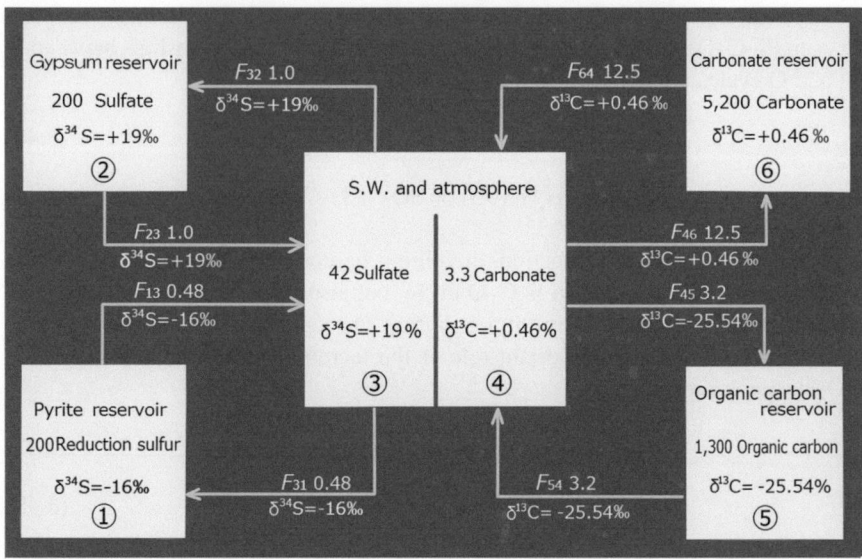

Fig. 5.11 Global S–C cycle steady state model (Garrels and Lerman 1984). Reservoir units: 10^{18} mol, Flux (F) units: 10^{18} mol my^{-1}. s.w. seawater

$$F_{ij} = k_{ij}M_i \tag{5.54}$$

(3) Paleozoic average sulfur isotopic composition ($\delta^{34}S_T$) is constant. For example,

$$\delta_1 S_1 + \delta_2 S_2 + \delta_3 S_3 = \delta S_T \tag{5.55}$$

If the mass of sulfur in the ocean remains constant, then changes in the masses of the sedimentary sulfates (reservoir 2) and sulfide (reservoir 1) must be of equal magnitude but opposite sign,

$$dS_2/dt = -dS_1/dt \tag{5.56}$$

The rate of change in the isotopic composition of sulfur in the ocean is a balance between the isotopic compositions of the input and removal fluxes, as given by the following equation,

$$d\delta_3/dt = (1/S_3)\{k_{13}\delta_1 S_1 + k_{23}\delta_2 S_2 - \delta_3 F_{32} - (\delta_3 - \alpha_s)F_{31}\} \tag{5.57}$$

where α_s is sulfur isotopic fractionation factor for biological reduction of sulfur from sulfate to sulfide which is 35 ‰.

These equations are converted into

$$dS_2/dt = F_{32} - k_{23}S_2 \tag{5.58}$$

$$d\delta_2/dt = \{(1/S_2)(-\delta_2 dS_2/dt - d(\delta_1 S_1)/dt - S_3 d\delta_3/dt)\} \tag{5.59}$$

Giving initial values (reservoir size, stable isotopic composition and flux), $\delta_1 S_1$, $\delta_2 S_2$, $\delta_3 S_3$, S_1, S_2, F_{31}, and F_{32} are obtained from above four differential equations. In a similar manner these parameter values for C cycle can be obtained. Temporal variation in sulfate (gypsum) reservoir size obtained from S cycle is in agreement with that from C cycle. This indicates that a reduction of carbonate carbon to organic carbon intimately relates to oxidation of sulfide to sulfate, implying that the ocean–atmosphere reservoir is small with respect to S, O, and C.

5.5 Global Geochemical Cycle-Mass Transfer Between Earth's Surface System and Interior System

As mentioned above, Garrels and Lerman (1984) and Berner (1987) solved the S–C–O cycle in earth's surface system (hydrosphere–atmosphere–crust system). However, earth's surface system is open to earth's interior (mantle). Mass transfer between hydrosphere, atmosphere, crust and mantle that is called global geochemical cycle will be considered below (Fig. 5.12).

The flux of hydrothermal solution at mid-oceanic ridges is estimated as 1.0×10^{23} g my^{-1} (Holland 1978). H$_2$S concentration of hydrothermal solution originated from seawater is 10^{-2}–10^{-3} mol kg^{-1}H$_2$O. Using these values, we obtain 10^{17}–10^{18} mol my^{-1} as H$_2$S flux from mid-oceanic ridges. Sulfur flux

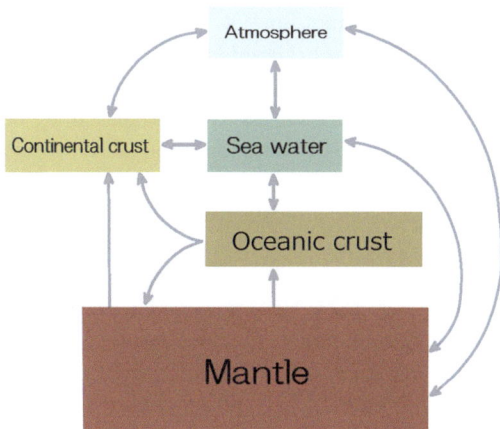

Fig. 5.12 Global geochemical cycle

between seawater and gypsum (evaporite) and between seawater and pyrite formed associated with sediment and biological activity is 1.0×10^{18} mol my^{-1} and 0.48×10^{18} mol my^{-1}, respectively (Garrels and Lerman 1984). These fluxes are not different from the hydrothermal flux. Sulfur in hydrothermal solution comes to ocean and atmosphere. This indicates that earth's surface system (ocean, atmosphere) is not closed to mantle.

5.5.1 Global Carbon Cycle

Input of CO_2 from mantle to ocean and atmosphere occurs by the degassing of volcanic gas and hydrothermal solution at arc-back arc and mid-oceanic ridges (Shikazono 2003). CO_2 input from atmosphere · ocean reservoir to mantle occurs by the subduction of sediments overlying oceanic plate.

Carbon fluxes between ocean · atmosphere reservoir and carbonate, and atmosphere and organic carbon are estimated as 12.5×10^{18} mol my^{-1} and 3.2×10^{18} mol my^{-1}, respectively. Hydrothermal CO_2 flux from mid-oceanic ridges is estimated to be $(1-2) \times 10^{18}$ mol my^{-1} from rate of seawater and hydrothermal solution cycling at mid-oceanic ridges and CO_2 concentration of hydrothermal solution. This flux is smaller than that between ocean–atmosphere reservoir and carbonate reservoir. CO_2 flux by hydrothermal solution associated with back arc volcanism today is estimated as which is higher or similar to hydrothermal CO_2 flux from mid-oceanic ridges (Shikazono 2003).

CO_2 flux by volcanic gas takes an important role in global carbon cycle. Gerlach (1989) estimated CO_2 flux by volcanic gas from mid-oceanic ridges today to be $(0.23-0.86) \times 10^{18}$ mol my^{-1}. Therefore, a summation of hydrothermal and volcanic gas CO_2 flux from mid-oceanic ridges is ca. 2×10^{18} mol my^{-1}. This carbon input by hydrothermal and volcanic gas changes to carbonate and organic carbon in atmosphere · ocean system. The amounts of carbonate and organic carbon are $5,200 \times 10^{18}$ mol and $1,300 \times 10^{18}$ mol, respectively and total amount is $6,500 \times 10^{18}$ mol (Garrels and Lerman 1984). If $6,500 \times 10^{18}$ mol is divided by carbon flux from mantle (2×10^{18} mol my^{-1}), we obtain 32×10^8 year. This may suggest that carbon has been degassed to ocean·atmosphere system gradually through 32×10^8 year, and this may support Rubey's continuous degassing model (Rubey 1951). However, it is widely accepted that carbon regases to mantle and degases to atmosphere–ocean system by diagenesis, igneous activity and metamorphic activity, related to plate subduction. If flux from mantle is nearly same to subduction flux to mantle, it is considered that most of carbon in earth's surface system has been degassed at early stage of earth's history.

Subduction carbon flux is estimated as follows. Carbon flux by subduction of altered oceanic crust and sediments is 1.7×10^{18} mol my^{-1} and 3.7×10^{18} mol my^{-1}, respectively. $\delta^{13}C$ of altered basalt is -7 ‰ which is same to mantle value. Therefore, only the flux by sediments to mantle is regarded as subduction carbon flux to mantle.

Carbon flux by volcanic gas from island arc and by hydrothermal solution from island arc and back arc is estimated as 0.15×10^{18} mol my^{-1}, and $(1-6) \times 10^{17}$ mol my^{-1}, respectively (Shikazono 1995). These fluxes are smaller than subduction flux (3.7×10^{18} mol my^{-1}). Therefore, it is considered that most carbon subducts to mantle. This subduction flux ($(3-4) \times 10^{18}$ mol my^{-1}) seems larger than hydrothermal and volcanic gas flux from mid-oceanic ridge ($(1-2) \times 10^{18}$ mol my^{-1}). Therefore, it seems likely that the amount of carbon in mantle reservoir increases with time and the large amounts of carbon stored in mantle may degass intermittently associated with superplume activity (Shikazono 1995).

It is inferred that long-term global carbon cycle including subduction, mantle storage and superplume activity is in steady state, indicating that continuous and gradual degassing is unlikely.

Residence time of carbon in mantle is estimated as $(1.5-2.0) \times 10^{8}$ year from subduction flux to mantle and initial amount of carbon in mantle (6×10^{21} mol) (Holland 1978). The estimated residence time indicates that most of carbon in mantle is recycled carbon. There is a question whether subducting carbon flux is same to mantle carbon degassing flux at mid-oceanic ridges. Subducting carbon is composed of carbonate carbon and organic carbon with the ratio of carbonate carbon:organic carbon = $5,200 \times 10^{18}$:$1,300 \times 10^{18}$. $\delta^{13}C$ of carbonate carbon and organic carbon is 1 ‰ and -25 ‰, respectively. Thus, average $\delta^{13}C$ of subducting carbon is -4 ‰ which is nearly same to $\delta^{13}C$ of mantle carbon and of hydrothermal solution from back arc basin (Shikazono et al. 1995). Thus, it is likely that carbon in hydrothermal solution and volcanic gas emitting from mid-oceanic ridges is mostly of subduction origin. It is obvious that plate subducts to upper mantle, but it may be possible that it goes to lower mantle and core. However, interactions between subduction plate, mantle and core are not well understood. Fluxes between these reservoirs have been estimated. Due to the subduction of plate earth's surface materials transfer to mantle, resulting to heterogeneous chemical and isotopic distributions in mantle. The causes for the mantle heterogeneity are subduction of plate, material transport from earth's interior (lower mantle, core) to earth's surface and heterogeneity of mantle formed at early stage of earth's history.

Figure 5.13 demonstrates the variations of ϵ_{Nd} and $^{87}Sr/^{86}Sr$ of mantle materials (MORB, ocean islands basalt, EM1, EM2, diamond, enriched mantle) where ϵ_{Nd} is defined as $\{(^{143}Nd/^{144}Nd)_{samp}/(^{143}Nd/^{144}Nd)_{chondrite} - 1\} \times 10^4$. ^{143}Nd generates by radiogenic decay of ^{147}Sm, while ^{144}Nd is stable isotope. It is obvious in Fig. 5.13 that ϵ_{Nd} and $^{87}Sr/^{86}Sr$ of mantle are lower than those of crust and are homogenously distributed. If we accept the above consideration on global carbon cycle, the change in atmospheric O_2 concentration obtained by Berner (1987) has to be reconsidered. It seems likely that high atmospheric O_2 content at Permian-Carboniferous age (ca. 3×10^6 years ago) calculated by Berner (1987) is probably not correct. Berner (1987) inferred that atmospheric O_2 concentration changes depending on the burial rate of organic matter. If the burial rate is fast, O_2 is consumed by the reaction

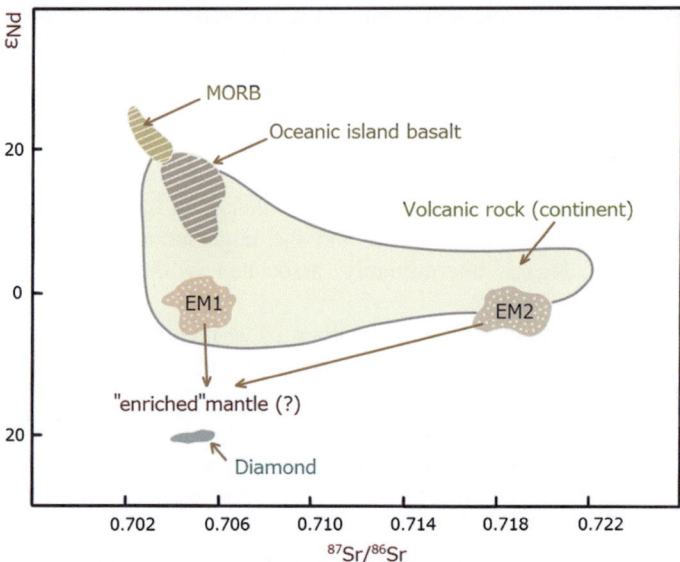

Fig. 5.13 ε_{Nd} and $^{87}Sr/^{86}Sr$ values of volcanic rocks and mantle material (diamond) (Silver and Carlson 1988)

$$CH_2O + O_2 \rightarrow CO_2 + H_2O \quad (5.60)$$

However, the burial rate is slow and reaction (5.60) does not proceed, resulting to high atmospheric O_2 concentration. In a real system, more complicated reaction occurs. That is expressed as

$$2Fe_2O_3 + 16Ca^{2+} + 16HCO_3^- + 8SO_4^{2-}$$
$$\rightarrow 4FeS_2 + 16CaCO_3 + 8H_2O + 15O_2 \quad (5.61)$$

If pyrite and calcite form during diagenesis according to reaction (5.61), O_2 generates. However decompositions of pyrite and calcite by weathering cause lower atmospheric O_2 concentration. Above reaction implies that atmospheric O_2 concentration depends not only on O_2 cycle but also on S, Ca and Fe cycles. Sr incorporates into carbonate. Therefore, O_2 concentration also depends on Sr cycle.

Considerable efforts to elucidate carbon cycle in atmosphere–ocean–solid earth system have been carried out since the pioneer work by Rubey (1951). Representative models on carbon cycle previously proposed are illustrated in Fig. 5.14 (Shikazono 1995). They are the following (1)–(4).

(1) Juvenile CO_2 has been degassing continuously from earth's interior to atmosphere–ocean system.
(2) Carbon has been cycling in earth's surface environment without plate subduction.

5.5 Global Geochemical Cycle-Mass Transfer Between Earth's Surface... 161

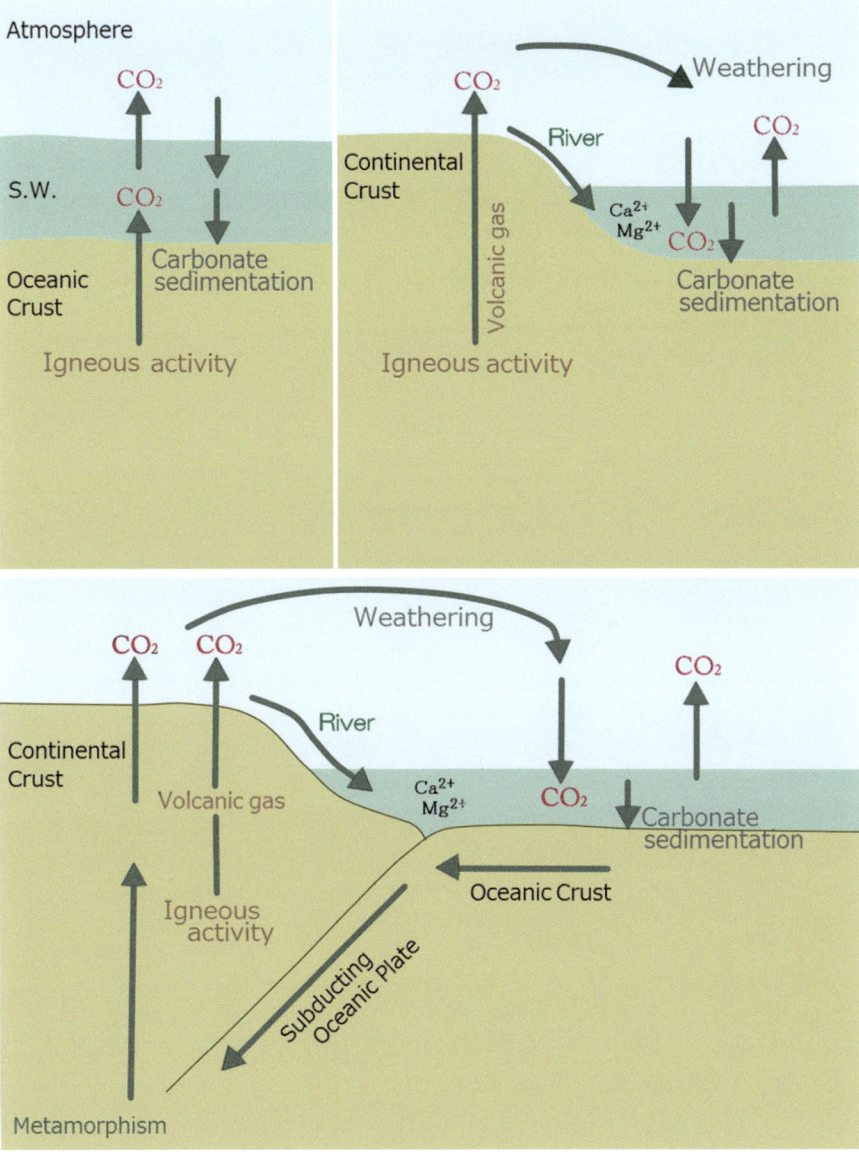

Fig. 5.14 Representative models of carbon cycle previously proposed (Shikazono 1997)

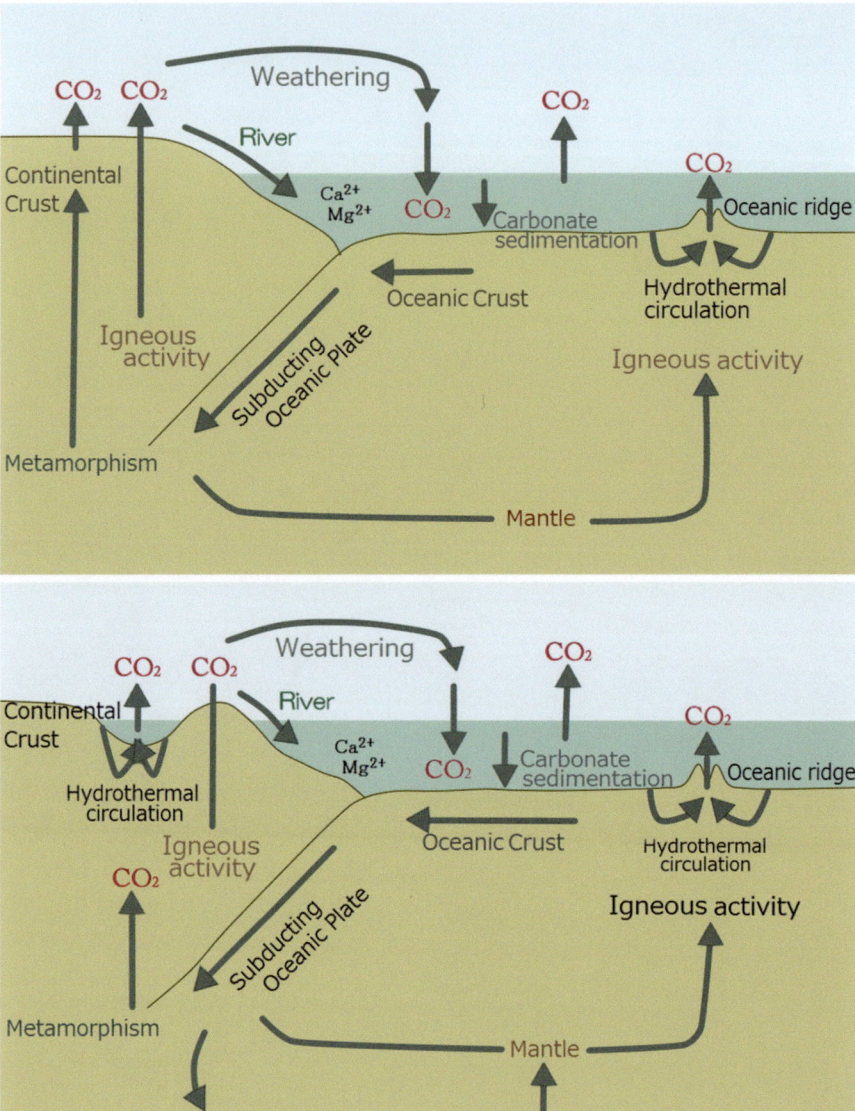

Fig. 5.14 (continued)

(3) Carbon has been cycling in earth's surface environment with plate subduction.

  (3-a) Subduction flux is same to degassing flux from subduction zone (e.g., BLAG model; Berner et al. 1983).
  (3-b) Subduction flux to mantle is same to degassing flux from mid-oceanic ridges.

(4) Subduction flux to mantle has been different from degassing fluxes from subduction zone and mid-oceanic ridges.

(1) takes into account degassing of CO_2 from earth's interior (mantle) and degassed CO_2 in atmosphere and ocean is thought to be taken as carbonates in ocean. Rubey's degassing model (Rubey 1951) and Hart's atmosphere evolution model (Hart 1978) correspond to this model. Rubey's model and Hart's model are different concerning the rate of degassing from earth's interior. However, both models consider that degassed gases do not return to earth's interior. It is discussed in detail in Matsui (1993) from various points of view that these models are not be widely accepted. After the formation of continents, Ca^{2+} and Mg^{2+} dissolved by weathering are transported by river water to ocean, and Ca^{2+} and Mg^{2+} combine with HCO_3^- resulting to precipitation of carbonates. This process causes reduction of atmospheric CO_2. This is negative CO_2 feedback by Walker et al. (1981).

(2) is silicate-carbonate model taking into account this negative feedback.

(3-a) model (BLAG model) is silicate-carbonate model and it considers carbon cycle between atmosphere, ocean and crust, taking into account plate tectonics. But this model does not consider carbon cycle between atmosphere, ocean and mantle. However, atmosphere–ocean–crust system is not closed to mantle with regard to carbon and carbon is continuously cycling between atmosphere–ocean and mantle. Javoy (1988) regarded that subducting carbon flux is same to carbon flux from mid-oceanic ridges. However, as mentioned above, these fluxes are different with each other.

(4) was proposed by the author (Shikazono 1995). According to this model, subducted CO_2 is stored in mantle and it returns to ocean·atmosphere system associated with superplume activity.

5.5.2 Global Sulfur (S) Cycle

The effect of volcanic gas on global sulfur cycle is important. S concentration of volcanic gas from island arc is smaller than CO_2 concentration. CO_2 concentration of hot spot volcanic gas is similar to that of island arc volcanic gas, but amount of volcanic gas emission from hot spot volcanic activity (e.g., Hawaii) is small (Table 5.3). S concentration of volcanic gas from mid-oceanic ridges is similar to C concentration. Emission of S and C by volcanic gas from mid-oceanic ridge is small because solubility of gas into magma is high under high pressure condition at deep sea depth (3,000–4,000 m). Sulfur in basalt transfers to hydrothermal solution

Table 5.3 Rate of cycling between mantle and crust (Kusakabe 1990)

	Rate of volcanic rocks production and subduction (10^{15} gy^{-1})	Concentration			Flux		
		H_2O (wt %)	CO_2 (wt %)	Cl (ppm)	H_2O (10^{14} gy^{-1})	CO_2 (10^{14} gy^{-1})	Cl (10^{12} gy^{-1})
Mid-oceanic ridge	56	0.2	0.04	48	1.1	0.2	2.7
Hot spot	4	0.3	0.02	290	0.1	0.01	1.3
Island arc volcanic rocks	5	2	–	900	1.0	–	4.5
	13	1	0.05	–	1.4	0.07	–
Altered oceanic crust	−60	1.5	0.1	50	−8.8	−0.6	−2.9
Sediments	−1.3	5	12	1,200	−0.7	−1.6	−1.6

during the circulation of seawater and hydrothermal solution at mid-oceanic ridges. H_2S concentration in hydrothermal solution is estimated to be 1/10 of CO_2 concentration from basalt–seawater interaction experiments at elevated temperature. Assuming that H_2S concentration is 1/10 of CO_2 concentration, H_2S flux from mantle through hydrothermal solution is estimated to be $(1–2) \times 10^{17}$ mol my^{-1}.

S flux by subduction will be estimated below. Using both S content of subducting altered basalt and sediments is as 0.1 wt% both and subducting rate of altered basalt and sediments as 60×10^{15} g year^{-1} and 1.3×10^{15} g year^{-1}, respectively, subducting S flux by altered basalt and sediments is estimated to be 2×10^{18} mol my^{-1} and 0.04×10^{18} mol my^{-1}, respectively. Sulfur in altered basalt is mostly of mantle origin. However, 20 % of total sulfur in altered basalt is derived from seawater (Kawahata and Shikazono 1988). Seawater sulfate is reduced to H_2S by the reaction with iron in sediments to form pyrite. Seawater sulfate is partially fixed as pyrite in altered basalt. Flux of subducting sulfur in pyrite is 4×10^{17} mol my^{-1}. Subducting flux of sulfur of seawater origin is that of sulfur in sediments and in pyrite in altered basalt $= 4.4 \times 10^{17}$ mol my^{-1}. Degassing sulfur flux by volcanic gas is reported as $(1.5–5) \times 10^{17}$ mol my^{-1} (Wolery and Sleep 1976) that is similar to or smaller than subduction sulfur flux. In smaller case, sulfur by subduction is stored in mantle and a part of stored sulfur may emit associated with volcanic activities at mid-oceanic ridges and superplume activities. Above discussion suggests that the balance between mantle and ocean has been maintained in long-term global sulfur cycle supports the view that continuous degassing of juvenile sulfur from mantle has not occurred, indicating that Rubey's continuous degassing model could be denied.

$\delta^{34}S$ of subducting sulfur gives a constraint on the global sulfur cycle. Using $\delta^{34}S$ of altered basalt (+4 ‰), $\delta^{34}S$ of sediments (−20 ‰), and the sulfur content of altered basalt vs. that of sediments = 60:1.3, we obtain average $\delta^{34}S$ of subducting sulfur = +1.5 ‰ that is similar to mantle $\delta^{34}S$. This suggests that mantle is contaminated by seawater sulfur transported by subduction. This supports above-consideration.

5.6 Geochemical Cycle of Minor Elements

5.6.1 Arsenic (As)

Recently, it has been clarified that hydrothermal solution from back arc basin and island arc contains considerable amounts of As. Thus, it is worthy to consider the As flux by hydrothermal solution and volcanic gas to seawater (Shikazono 1993). For example, the concentration of hydrothermal solution from Lau back arc basin is very high (6,000–11,000 mmol kg^{-1} H_2O) (Fouquet et al. 1991). Assuming that As concentration of hydrothermal solution from back arc basin at middle

Miocene Sea of Japan and circulation rate of seawater and hydrothermal solution at Green tuff region is 8×10^{14} g year^{-1} (Shikazono 1994), As flux by cycling of seawater and hydrothermal solution is estimated as $(1.1–5.3) \times 10^{13}$ mol my^{-1}. If the area of back arc basin in Circum Pacific region was 20–30 times of Green tuff region in Japan, total flux is large $(2–16) \times 10^{14}$ mol my^{-1}, compared with riverine flux today (1.0×10^{15} mol my^{-1}). Volcanic eruption rate at island arc is 0.75 km^3 year^{-1} that is 5–7 times of that at back arc basin. Combining this rate with As concentration of hydrothermal solution from island arc, hydrothermal As flux is estimated as $(2.5–61) \times 10^{14}$ mol my^{-1}. It is clear from above estimation that hydrothermal As flux from arc-back arc system is very large. As flux from the atmosphere to ocean is 3.5×10^{13} mol my^{-1} (Walsh et al. 1979) that is negligibly small, compared with the hydrothermal As flux to ocean. As concentration of hydrothermal solution at mid-oceanic ridges is 0.01–0.02 ppm that is very small, compared with hydrothermal solution at back arc basins. Hydrothermal As flux at mid-oceanic ridges is $(1.1–2.1) \times 10^{13}$ mol my^{-1} that is 1/10–1/300 of As flux at arc-back arc basins.

Arsenic(As) in ocean is mainly removed by formation of pyrite in marine sediments. The production rate of sulfur in pyrite is 3.3×10^{18} mol my^{-1} (2.5×10^{20} g my^{-1}) (Holland 1978). As/S ratio of pyrite in sediments previously reported is $(8.7 \pm 3) \times 10^{4}$ (Huerta-Diaz and Morse 1992). Thus, As sink by pyrite is $(1.7–3.9) \times 10^{15}$ mol my^{-1}. This flux seems to be not different from As input to ocean (($1.6–8.1) \times 10^{15}$ mol my^{-1} (Table 5.3). As concentration of ocean is considered to be controlled by hydrothermal input, riverine input and pyrite output. Fluxes by volcanic gas from atmosphere and by weathering of ocean-floor basalt are small, compared with hydrothermal, riverine and pyrite As fluxes. Residence time of As in seawater is estimated as the amount of As in seawater (4.2×10^{15} g) divided by As input to seawater ($(1.6–8.1) \times 10^{15}$ mol my^{-1} which is equal to $(1.7–3.8) \times 10^{4}$ year. This is shorter than previously estimated one (10^5 year by Holland 1978). Subducting sulfur flux is estimated to be 6.1×10^{19} g my^{-1} from S contents of altered basalt and sediments (~0.1 wt%) (Kawahata and Shikazono 1988) and subducting rates of altered basalt (60×10^{21} g my^{-1}) and sediments (1.3×10^{21} g my^{-1}) (Kusakabe 1990). As/S of pyrite is $(8.7 \pm 3) \times 10^{-14}$. Thus, subducting As flux is $(0.7–14.7) \times 10^{12}$ mol my^{-1} (($5.3–11) \times 10^{14}$ g my^{-1}). This is similar to or smaller than hydrothermal As flux from island arc-back arc system, but it is larger than volcanic gas As flux (3.5×10^{13} mol my^{-1}). These estimates indicate that most of subducting As return to atmosphere–ocean system and do not transfer to mantle. It is also likely that As in crust goes to atmosphere–ocean system by hydrothermal solution and volcanic gas.

Output from ocean and input of As to ocean are summarized in Table 5.4.

It has been cited that As concentration in atmosphere is largely influenced by human activity from the analytical data on aerosol which contains significant amounts of Pb, Hg, Cu, As etc. However, it was found recently that volcanic emission contains large amounts of these element, indicating volcanic activity

Table 5.4 Geochemical balance of arsenic for ocean and subduction flux (g/year) (Shikazono 1993)

Input flux to ocean	
(1) Riverine flux	7.8×10^{10}
(2) Hydrothermal flux:	
island arc-back-arc	$(0.2–5.2) \times 10^{10}$
basin axis of mid-ocean ridge	$(0.8–1.6) \times 10^{11}$
(3) Volcanic gas	2.8×10^{9} (max.)
(4) Atmosphere	2.6×10^{9}
(5) Weathering of ocean floor of basalt	$(2.7–9) \times 10^{8}$
Total:	$(1.0–6.1) \times 10^{11}$
Output flux from ocean	
(6) Sedimentation (formation of pyrite)	$(1.3–2.9) \times 10^{11}$
(7) Atmosphere	1.4×10^{8}
Total:	$(1.3–2.9) \times 10^{11}$
Subduction flux	$(4.0–8.2) \times 10^{10}$

influences As and the base metal elements concentrations in atmosphere. This consideration is consistent with the estimate that the influence of hydrothermal flux to the atmospheric As concentration is large.

5.6.2 Boron (B)

Hydrothermal B flux from island arc and back arc basin will be obtained in a similar manner mentioned above. B concentration of hydrothermal solution issuing from Lau back arc basin and the other island arc and back arc basins is in a range of 770–870 μm (Fouquet et al. 1991; Gamo 1995). From this concentration and circulation rate of seawater and hydrothermal solution at back arc basins today, we obtain $(2.3–12) \times 10^{14}$ mol my^{-1} as hydrothermal B flux from back arc basins. However, eruption rate of island arc volcanic rocks is 0.75 km^3 year^{-1} that is significantly higher than 0.1 km^3 year^{-1} of that of back arc volcanic rocks. Thus, it is likely that hydrothermal flux from island arc is one magnitude of higher than hydrothermal B flux from back arc basin ($(2.3–12) \times 10^{14}$ mol my^{-1}). Thus, this flux seems important because this cannot be neglected compared with riverine B flux (4.3×10^{16} mol my^{-1}). Most of subducting B degases to crust–atmosphere–ocean system by hydrothermal solution and volcanic gas, indicating that the flux of B subducting to mantle is very small. This is supported by the positive correlation between B content and ^{10}Be content of island arc volcanic rocks (Morris et al. 1990; Leeman et al. 1994). ^{10}Be is induced by cosmic ray in the atmosphere and goes to ocean to ocean floor sediments subducting to deeper parts where magma generates.

Existence of ^{10}Be in island arc volcanic rocks whose half life time is ca. 1.5 my clearly indicates that island arc volcanic rocks are influenced by seafloor sediments.

5.6.3 Barium (Ba)

Average Ba concentration of hydrothermal solution from back arc basins is 5 ppm. Ba flux is obtained from 5 ppm × (2.5–13) × 10^{15} g my^{-1} (cycling rate of seawater and hydrothermal solution at back arc basins) that is (1.1–6.5) × 10^6 g year^{-1}. Eruption rate of island arc volcanic rocks is 75 times of that of back arc volcanic rocks (Katsui and Nakamura 1979). Thus, hydrothermal Ba flux from island arc-back arc systems is estimated as (8–47) × 10^{15} mol my^{-1}. Cycling rate of seawater and hydrothermal solution at middle Miocene back arc basins when Kuroko deposits formed at Sea of Japan is higher than that today, suggesting the hydrothermal flux at middle Miocene was higher than that today. Riverine Ba flux today is 1.0×10^{16} mol my^{-1} that is similar to hydrothermal Ba flux today. Most of Ba derived by hydrothermal solution to ocean precipitate as barite (BaSO$_4$) near the hydrothermal vent. Ba dispersed away from the vent is taken up by foraminifera. Dead foraminifera settles onto seafloor and dissolves in the sediments, resulting to the formation of barite nodule. The barite nodule may subduct with plate motion. Ba flux from mid-oceanic ridge hydrothermal solution is 4.0×10^{14} mol my^{-1} which is of same order of magnitude but smaller than the flux from island arc.

5.6.4 Other Ore Constituent Elements

In the above, we estimated hydrothermal flux of some elements mainly based on the data on chemical compositions of hydrothermal solution. However, large number of analytical data are not available and very few data on minor elements except the elements mentioned above have been obtained. Thus, we try to consider hydrothermal fluxes of minor elements based on analytical data on Kuroko ore deposits which have been formed at Miocene back arc basin (Sea of Japan).

Figure 5.15 shows the relationship between average chemical composition of Kuroko ore and that of granitic rocks which can be roughly regarded as average crustal composition. Kuroko ore contains appreciable amounts of Hg, Tl, Pb, Zn, Cu, As, Sb, Bi, Se, In, Ag, and Cd which occur as sulfides. Ba, Hg, As, Cu, Pb, Si, Ag, Au, Bi, Ga, Ge, Sb, Sr and Tl are more enriched in Kuroko ores, compared with mid-oceanic ridge hydrothermal ore deposits. Namely, hydrothermal solution responsible for Kuroko deposits contains appreciable amounts of soft elements by Pearson (1968) and volatile elements. Figure 5.15 suggests that hydrothermal fluxes of the elements enriched into Kuroko deposits (As, Hg, Ba, Cu, Pb, Si, Ag, Au, Bi etc.) from island arc-back arc basins are large. However, this is not obvious because most of the amounts of elements in hydrothermal solution do not precipitate and fix as minerals. Efficiency of precipitation of elements differs in different mineral and fractionation of elements between hydrothermal solution and minerals occur. Figure 5.16 shows the relationship

5.6 Geochemical Cycle of Minor Elements

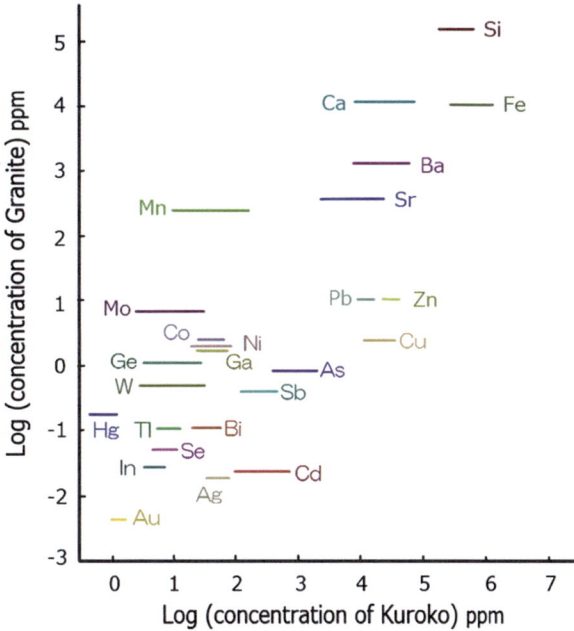

Fig. 5.15 Average chemical compositions of Kuroko ore and granite (Kajiwara 1983; Shikazono 1988)

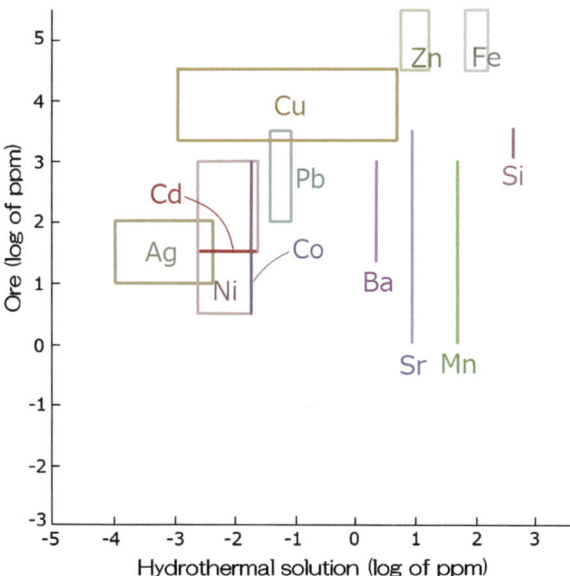

Fig. 5.16 Chemical compositions of ridge hydrothermal solution and ore deposits (Shikazono 1988)

between chemical composition of mid-oceanic ridge hydrothermal solution and that of ores. It is found in Fig. 5.16 that precipitations of Fe, Mn, and Ba are not efficient and they disperse into ocean. In contrast Cu, Zn, Pb, Ag, Au, Bi, Hg, Sb and As tend to be fixed as sulfides in ore deposits. This tendency is consistent with the data shown in Fig. 5.15.

Cited Literature

Barron EL, Sloan JL II, Harrison CGA (1978) Palaeogeogr Palaeoclim Palaeoecol 30:17–40
Berner RA (1987) Am J Sci 287:177–196
Berner RA (1991) Am J Sci 291:339–376
Berner RA (1994) Am J Sci 294:56–91
Berner RA, Kothavala Z (2001) Am J Sci 301:184–204
Berner RA, Lasaga AC, Garrels RM (1983) Am J Sci 283:641–683
Bischoff JL, Radke AS, Rosenbauer RJ (1981) Econ Geol 76:659–676
Fouquet Y, Stackelberg U, Von Charlou JL, Donval JP, Foucher JP, Erzinger J, Herzig P, Muhe R, Wiedicke M, Soahai S, Whitechurch H (1991) Geology 19:303–306
Gamo T (1995) In: Sakai H, Nozaki Y (eds) Biogeochemical processes and ocean flux in the western pacific. Terra Scientific Publisher, Tokyo, pp 425–451
Garrels RM, Lerman A (1984) Am J Sci 284:989–1009
Garrels RM, Perry EA Jr (1974) In: Goldberg ED (ed) The sea, vol 5. Wiley, New York, Chap. 9
Gerlach TM (1989) J Volcanol Geotherm Res 39:221–232
Hart MH (1978) Icarus 33:23–39
Huerta-Diaz MA, Morse JW (1992) Geochim Cosmochim Acta 56:2681–2702
Ivanov MV, Grinenko VA, Rabinovich AP (1983) In: Ivanov MV, Freney JR (eds) The global biogeochemical sulphur cycle, SCOPE 19. Wiley, Chichester, pp 331–356
Javoy M (1988) Chem Geol 70:39
Kashiwagi H, Shikazono N (2003) Palaeogeogr Palaeoclim Palaeoecol 199:167–185
Kashiwagi H, Ogawa Y, Shikazono N (2008) Palaeogeogr Palaeoclim Palaeoecol 270:139–149
Katsui Y, Nakamura K (1979) Chikyukagaku. Iwanami Shoten 7:195–213 (in Japanese)
Kawahata H, Shikazono N (1988) Can Min 26:555–565
Kennet JP, McBirney AR, Thunell RC (1977) J Volc Geotherm Res 2:145–163
Kramer JR (1965) Geochim Cosmochim Acta 29:921–945
Kusakabe M (1990) Science (Kagaku) 60:711–712 (in Japanese)
Lasaga AC, Berner RA, Garrels RM (1981) In: Sundquist ET, Broecker WS (eds) The carbon cycle and atmospheric CO_2; Natural variations Archean to present. Am Geophysical Union, Washington DC, pp 397–411
Leeman WP, Carr MJ, Morris JD (1994) Geochim Cosmochim Acta 58:149–168
Lerman A (1971) Adv Chem Ser 106:30–76
Lerman A, Mackenzie FT, Garrels RM (1975) Geol Soc Am Mem 142:205–218
Lovelock JE, Maggs RJ, Rasmussen RA (1972) Nature 237:452–453
Matsui T (1993) J Geogr 102:633–644 (in Japanese with English abstract)
Morris G, Leema WP, Tera F (1990) Nature 344:31–36
Pagani M, Arthur MA, Freeman KH (1999) Palaeoceanogr 14:273–292
Pagani M, Lemarchand D, Spivak A, Gaillardet J (2005) Geochim Cosmochim Acta 69:953–961
Pearson RG (1968) J Am Chem Educ 45:581–587; 643–648
Pearson PN, Palmer MR (2000) Nature 406:695–699
Pitman WC (1978) Geol Soc Am Bull 89:1289–1403
Rubey WW (1951) Geol Soc Am Bull 62:1111–1148
Shikazono N (1993) Chikyukagaku (Geochem) 27:135–139 (in Japanese with English abstract)

Shikazono N (1994) Island Arc 3:59–65
Shikazono N (1995) Science (Kagaku) 65:324–332 (in Japanese)
Shikazono N, Kashiwagi H (2007) Netsusokutei 34:68–76 (in Japanese with English abstract)
Shikazono N, Shimizu M, Utada M (1995) Appl Geochem 10:621–642
Walker JCG, Hays PB, Kasting JF (1981) J Geophys Res 86:9776–9782
Wolery TJ, Sleep NH (1976) J Geol 84:249–275

Further Reading

Chameides WL, Perdue EM (1997) Biogeochemical cycles a computer-interactive study of earth system science and global change. Oxford University Press Inc, Oxford
Ernst WG (2000) Earth systems. Cambridge University Press, Cambridge
Holland HD (1978) The chemistry of the atmosphere and oceans. Wiley, New York/Chichester/Brisbane/Toronto, p 351
Holland HD (1984) The chemical evolution of the atmosphere and oceans. Princeton University Press, Princeton
Jacobson MC, Charlson RJ, Rodhe H, Orians GH (2004) Earth system science. International Geophysics Series, vol 72. Elsevier/Academic
Kump LR, Kasting JR, Crane RG (1999) The earth system. Pearson Prentice Hall, Upper Saddle River, p 07458
Lasaga AC, Kirkpatrick RJ (eds) (1981) Kinetics of geochemical processes. Reviews in mineralogy (Am Min), vol 8
Lerman A (1979) Geochemical processes-water and sediment environments. Wiley, New York
Shikazono N (2003) Geochemical and tectonic evolution of arc-backarc hydrothermal systems. Elsevier, Amsterdam
Shikazono N (2010) Introduction to earth and planetary system science. Springer, New York

Chapter 6
Interaction Between Nature and Humans

In previous Chapters we dealt with mass transfer mechanism in water–rock interactions, geochemical modeling of water–rock interaction and global geochemical cycle in earth system. However, recent increase in the amount of various kinds of waste from humans influences significantly on the interactions in earth system, causing the change in environment, and giving negative feedback to humans. Thus, we cannot discuss and consider the mass transfer mechanism and global geochemical cycle in earth system without taking into account the anthropogenic influence on the natural system.

Several types of nature–humans interactions with respect to materials and energy include resources, waste and global environmental problems (Fig. 6.1). In this Chapter we focus on the global environmental and waste problems such as acid rain, CO_2 emission, underground CO_2 sequestration, geological disposal of high level nuclear waste, and water and soil pollutions.

6.1 Flux to the Atmosphere Due to Human Activity

Emissions of various kinds of gases caused by human activities (burning of fossil fuels, production of concrete, smelting etc.) occur. It is obvious that anthropogenic fluxes exceed natural fluxes for many elements (Table 6.1).

6.1.1 Carbon Dioxide (CO_2)

Natural and anthropogenic CO_2 fluxes are summarized in Table 6.2. It is obvious that CO_2 fluxes due to burning of fossil fuels and production of concrete are large, corresponding to ca. 7 % of carbon fluxes by photosynthesis and decomposition of organic matter. The natural fluxes such as precipitation of carbon, weathering of

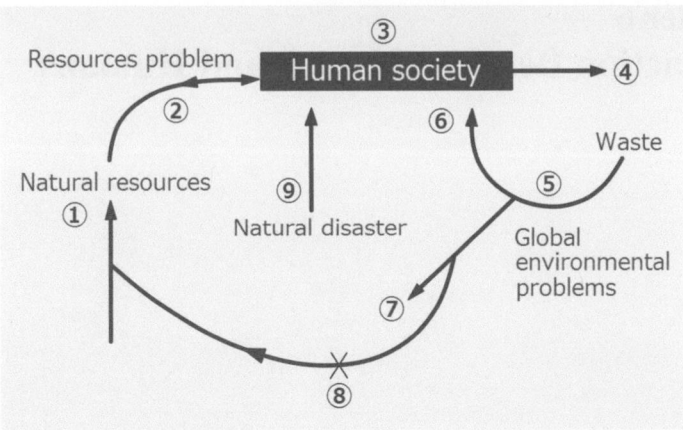

Fig. 6.1 Mass flow in nature–humans system (Shikazono 1994)

carbonate, formation of carbonate, and metamorphic and igneous degassing are negligibly small, compared with anthropogenic CO_2 flux.

Recently, anthropogenic CO_2 flux has been significantly increasing (Fig. 6.2). We can roughly estimate the anthropogenic CO_2 flux and atmospheric CO_2 concentration, if no CO_2 sink from atmosphere occurs. However, this estimate is not in agreement with actual atmospheric CO_2 concentration, indicating that some of CO_2 emitted are removing from the atmosphere to natural system (ocean, forest, desert, rocks). This CO_2 missing sink problem is a sort of enigmatic. However, a possible candidate for this sink is considered to be growing of forest in northern hemisphere by IPCC (1994). However, it is also proposed that the sink by desert and chemical weathering may take important role for causing a reduction of atmospheric CO_2 concentration (e.g., Tanaka 1994).

6.1.2 Sulfur (S)

Figures 6.3 and 6.4 show preindustrial and present-day circulation of S in earth's surface environment. Sulfur supply to the atmosphere by industrial activities (e.g., burning of fossil fuels, smelting) is 113×10^{12} g year^{-1} that is about eight times of flux by volcanism (14×10^{12} g year^{-1}) (Kimura 1989). Riverine sulfur flux to ocean is 208×10^{12} g year^{-1}. A half of this flux is considered to be of anthropogenic source (Holland 1978). Sulfur in environment (atmosphere, river water) is the element that is significantly affected by human activity, the greatest among elements. According to previous estimates most of sulfur in acid rain transfer to river water. However, acid rain containing sulfur reacts with soil and evaporite, leading to the formation of sulfate minerals and fixation of sulfur in soil. If we take into account the amount of sulfur fixation as sulfates in soil, previously obtained

6.1 Flux to the Atmosphere Due to Human Activity

Table 6.1 Estimate of emission to atmosphere from various sources (10^{18} g year^{-1}) (Nishimura 1991)

Element	Rock weathering	Volcanic eruption	Volcanic gas	Industrial activity	Coal, oil burning	Sum of rock weathering and volcanism	Sum of (?)	Contribution of anthropogenic condition (%)	Flux through atmosphere/Flux through river
Al	356,500	132,750	8.4	40,000	32,000	489,260	72,000	13	0.002
Ti	23,000	12,000	–	3,600	1,600	35,000	5,200	13	0.003
Sm	32	9	–	7	5	41	12	23	0.003
Fe	190,000	87,750	3.7	75,000	32,000	277,750	107,000	28	0.005
Mn	4,250	1,800	2.1	3,000	160	6,050	3,160	34	0.019
Co	40	30	0.04	24	20	70	44	39	0.018
Cr	500	84	0.005	650	290	584	940	62	0.042
V	500	150	0.05	1,000	1,100	650	2,100	76	0.079
Ni	200	83	0.0009	600	380	283	980	78	0.092
Sn	50	2.4	0.005	400	30	52	430	89	–
Cu	100	93	0.012	2,200	430	193	2,630	93	0.236
Cd	2.5	0.4	0.001	40	15	2.9	55	95	0.425
Zn	250	108	0.14	7,000	1,400	358	8,400	96	0.400
As	25	3	0.1	620	160	28	780	97	0.967
Se	3	1	0.13	50	90	4	140	97	1.111
Sb	9.5	0.3	0.013	200	180	19.8	380	97	0.340
Mo	10	1.4	0.02	100	410	11	510	98	0.443
Ag	0.5	0.1	0.0006	40	10	0.6	50	99	0.077
Hg	0.3	0.1	0.001	50	60	0.4	110	100	7.885
Pb	50	8.7	0.012	16,000	4,300	59	20,300	100	1.213

Table 6.2 Natural and anthropogenic CO_2 flux (Holland 1987) (C flux: unit $\times 10^5$ g year^{-1})

Transfer Between Reservoirs			
From	To	Process	Rate (in 10^{15} g/year)
7	1	Fossil fuel burning	4.2
8	1	Cement manufacture	0.7
7	1	Oxidation of elemental carbon	0.09 ± 0.02
6	7	Deposition of elemental carbon	0.09 ± 0.02
1	6	Net flux to carbonates	0.06 ± 0.04
8	6	Weathering of carbonates	0.16 ± 0.04
6	8	Deposition of carbonates	0.22 ± 0.04
7, 8, 9	1	Degassing due to metamorphic and igneous processes	0.09 ± 0.03

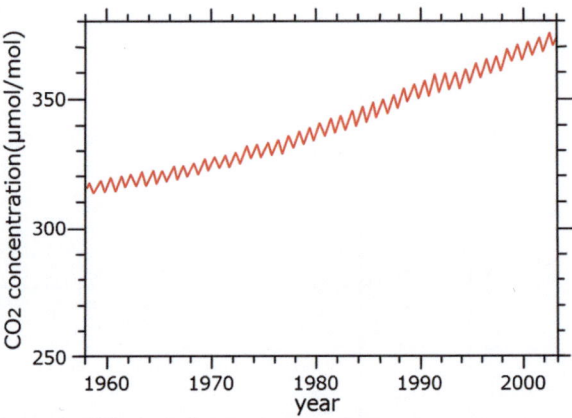

Fig. 6.2 Atmospheric CO_2 concentration in recent years based on direct determinations at Mauna Loa Observatory (Keeling and Whort 2003; Marini 2007)

anthropogenic contribution to riverine sulfur is considered to be overestimated. We have to be careful when we use previously estimated fluxes for the calculation on geochemical cycle and take for the processes that have not been considered in previous studies. Further, we have to take into account temporal change in fluxes. Sulfur flux has been studied by several researchers as well as carbon flux. However, the previously obtained sulfur flux differs in different researchers (Table 6.3).

6.1.3 Phosphorus (P)

Figure 6.5 shows short-term P cycle (Kimura 1989). Most of P is present in geosphere and hydrosphere, but it is small in amount in the atmosphere. Thus, the fluxes between atmosphere and geosphere and atmosphere and hydrosphere are small (Fig. 6.6). The amount of P in soil is large (($96-160) \times 10^{15}$ g). This is due to an increase in P flux as fertilizer. The amount of P in biosphere is 26×10^{14} g and $(5-12) \times 10^{13}$ g in terrestrial and marine organisms, respectively. Figure 6.6 shows six reservoir models for the global biogeochemical cycle of P.

6.1 Flux to the Atmosphere Due to Human Activity

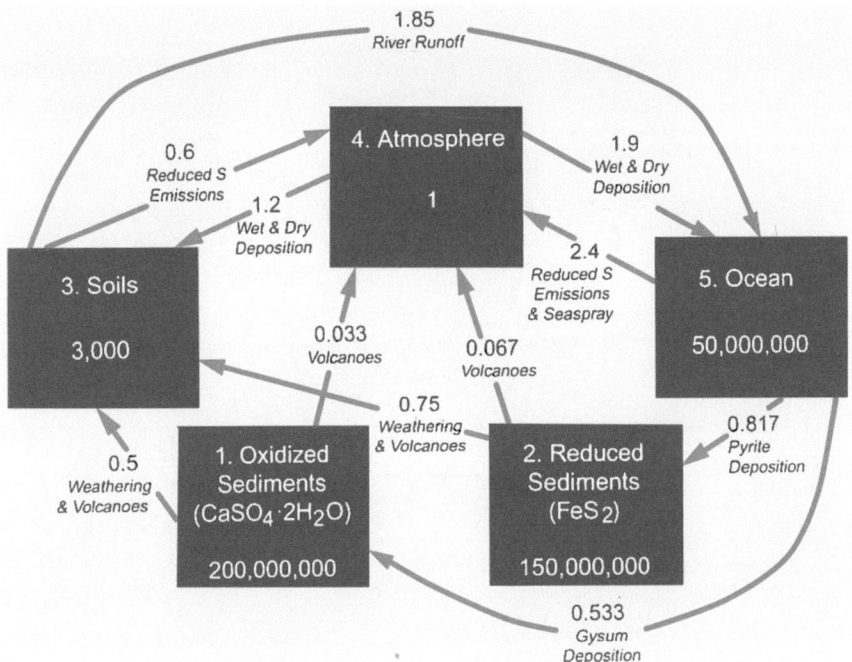

Fig. 6.3 The five-reservoir model for the preindustrial, steady-state S cycle with reservoirs for oxidized and reduced S sediments. Reservoir amounts are given in units of T moles and fluxes in T moles year^{-1} (Recall that 1 Tmole $= 1 \times 10^{12}$ mol) (Chameides and Perdue 1997)

6.1.4 Minor Elements

Anthropogenic minor element fluxes to atmosphere are summarized in Table 6.4 (Nishimura 1991). It is clear that anthropogenic fluxes of Pb, Hg, Ag, Mo, Cd, Cu, and Sn are greater than weathering + volcanic fluxes of these elements. The fluxes are mainly due to smelting of ore, and burning of oil, coal and natural gas (Appendix, Plate 34). In contrast, anthropogenic contribution for Al, Ti, Sn, Fe, Mn and Co are small compared with natural fluxes (weathering and volcanic flux). In recent years, chemical analyses of aerosol have been carried out, but the analytical data on minor elements are few and sources of these elements in aerosol are not clarified.

Table 6.1 shows fluxes for metals from mines, emission to atmosphere, transportation by rivers, and removal by rainwater. Most of anthropogenic metals transported to atmosphere are taken up by rainwater. However, this is unclear because fluxes to the atmosphere are not only by human activity but also by natural process such as volcanic activity and estimated proportion of these two processes has large uncertainty. Riverine fluxes to ocean are mostly larger than anthropogenic fluxes to atmosphere and fluxes by rainwater, but anthropogenic fluxes to atmosphere for toxic metals (Pb, Hg, Se, As, and Sb) are considerably large.

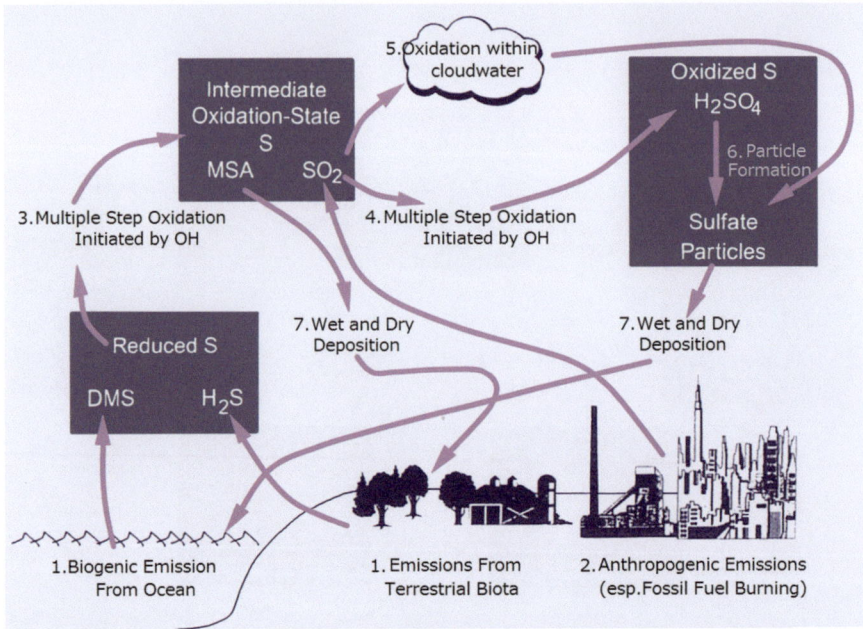

Fig. 6.4 Schematic illustration of the key pathways in the atmospheric cycle of S involving: (*1*) the natural emissions of reduced S compounds such as H_2S from terrestrial biota and dimethyl sulfide (CH_3SCH_3) from oceanic biota; (*2*) anthropogenic emissions of S compounds, principally SO_2; (*3*) the oxidation of reduced S compounds by OH and other photochemical oxidants leading to the production of intermediate oxidation state S compounds such as SO_2 and methane sulfonic acid (MSA); (*4*) the oxidation of these intermediate oxidation state compounds within the gas phase by OH-producing H_2SO_4 vapor; (*5*) the conversion of intermediate oxidation state compounds within liquid could droplets, which upon evaporation yield sulfate-containing particles; (*6*) the conversion of H_2SO_4 to sulfate-containing particles; and (*7*) the ultimate removal of S from the atmosphere by wet and dry deposition (Chameides and Perdue 1997)

Table 6.3 Previously obtained riverine sulfur flux to ocean

	Meybeck (1978)	Ivanov et al. (1983)	Husar and Husar (1985)
Number of river	40	–	54
Total runoff	3.74	4.24	4.1
Average S concentration	2.8	4.9	3.2
Natural riverine S flux	–	104	46–85
Anthropogenic riverine S flux	–	93	46–86
Total riverine flux	–	208	131–170

6.1 Flux to the Atmosphere Due to Human Activity

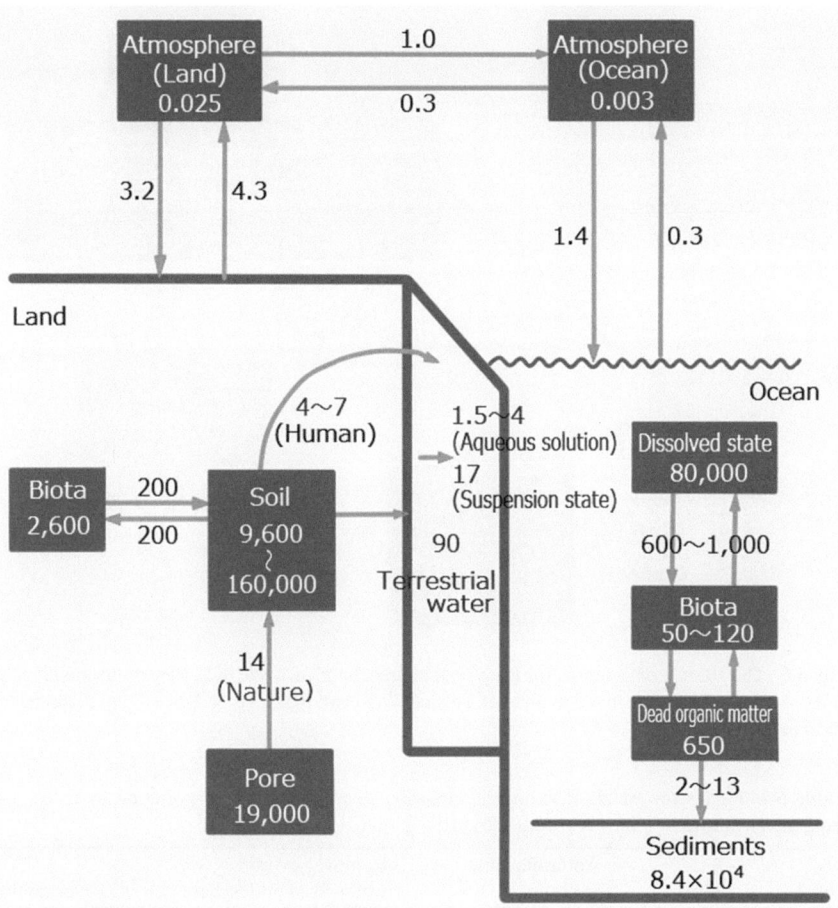

Fig. 6.5 Short-term P cycle and amount of P in earth's surface environment (Kimura 1989). Unit of flux is 10^6 t·year^{-1}

6.1.5 Geochemical Cycles of Pb, Cd and Hg Have Been Well Investigated Because of Their High Toxicity

Figure 6.7 shows Hg geochemical cycle at pre-industrial and industrial activity stages. It is clear that Hg emission to atmosphere and riverine Hg flux has increased due to human activity such as mining of Hg. It is considered that 2/3 of Hg fluxes to ocean and terrestrial environment is of anthropogenic source (Mason et al. 1994). It is important to know chemical state of Hg and reactions involving Hg in order to elucidate Hg cycle (Bunce 1991).

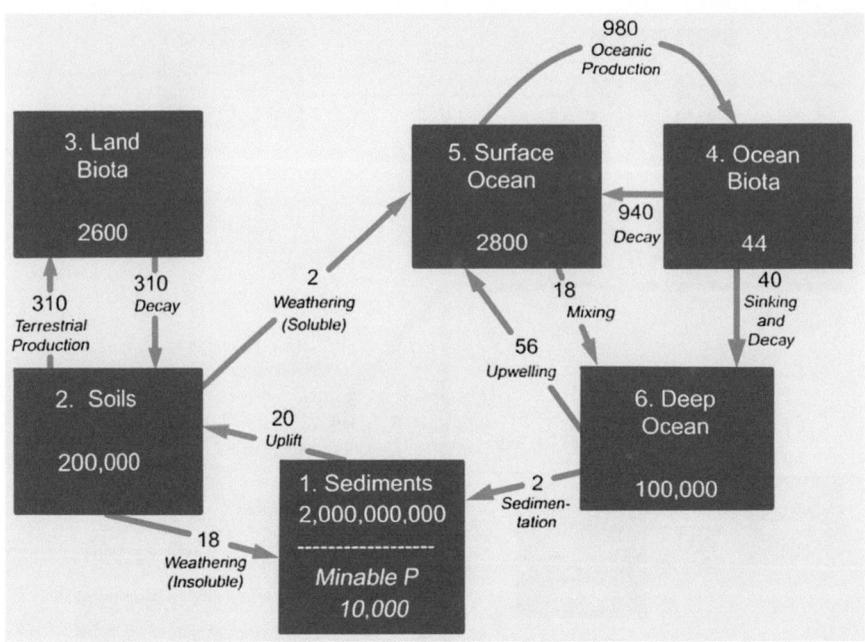

Fig. 6.6 The six-reservoir model for the global biogeochemical cycle of P. Reservoirs are given in units of Tg and fluxes are given in units of Tg year^{-1} (Recall that 1 Tg = 1 × 10^{12} g) (Chameides and Perdue 1997)

Table 6.4 Fluxes for metals from mines, emission to atmosphere, transportation by rivers and removal by rainwater (Garrels et al. 1975) (10^{12} g year^{-1})

Metal	Mine	Anthropogenic emission	Removal by rain water from atmosphere	Riverine flux
Pb	3	0.40	0.31	0.42
Cu	6	0.21	0.19	0.82
V	0.02	0.09	0.02	2.4
Ni	0.48	0.05	0.12	1.2
Cr	2	0.05	0.07	1.7
Sn	0.2	0.04	–	0.27
Cd	0.014	0.004	–	0.04
As	0.06	0.05	0.19	0.3
Hg	0.009	0.01	0.001	0.005
Zn	5	0.73	1.04	1.8
Se	0.005	0.009	0.03	0.02
Ag	0.01	0.003	–	0.03
Sb	0.07	0.03	0.03	0.09

6.1 Flux to the Atmosphere Due to Human Activity

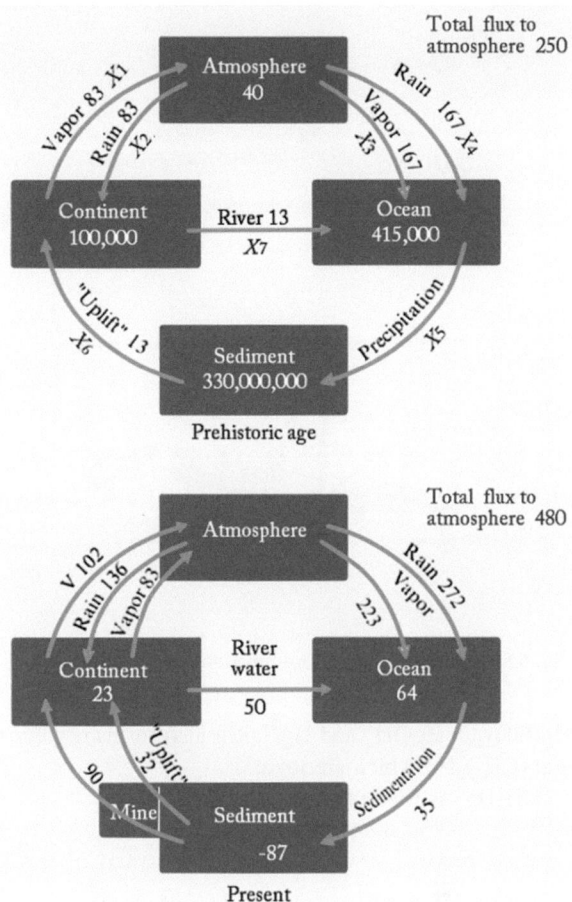

Fig. 6.7 Geochemical cycle of Hg at prehistoric age and present (Garrels et al. 1995). Unit of mass of reservoir is 10^8 Gt and that of flux is 10^8 Gt year^{-1}

HgS is taken from Hg mine. The following reaction occurs, if HgS is heated in air.

$$HgS + O_2 \xrightarrow{700\,C°} Hg(g) + SO_2 \tag{6.1}$$

Hg gas is emitted to the atmosphere by this reaction.

Hg is present as Hg^{2+} in aqueous solution. If Hg_2Cl_2 dissolves into aqueous solution, the following reaction occurs.

$$2Hg^+ \rightarrow Hg^{2+} + Hg \tag{6.2}$$

Hg is present in aqueous solution as various chemical states. Hg chloro- and thio complexes are dominant in aqueous solution with high Cl^- and HS^- concentrations. In the case of low concentration of Cl^- and HS^-, Hg is present as

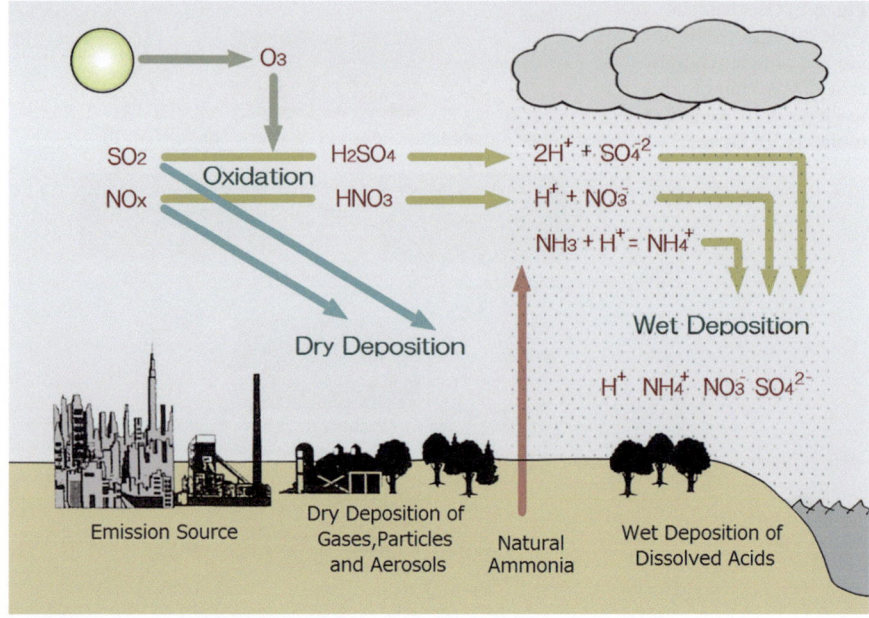

Fig. 6.8 Reactions causing for formation of acid rains

$Hg(OH)_2O$, $HgOH^+$ and Hg^{2+}. Hg also forms organic complexes such as CH_3Hg^+ and $(CH_3)_2Hg$ which are toxic.

CH_3Hg^+ is formed by the reaction

$$L_5Co\text{-}CH_3 + Hg \rightarrow L_5Co^+ + CH_3Hg^+ \qquad (6.3)$$

CH_3Hg^+ changes to CH_3HgCl and CH_3HgSCH_3 in marine organism.

6.2 Anthropogenic Fluxes to the Hydrosphere and Soils and Mass Transfer Mechanism

The hydrosphere is mostly composed of rainwater, river water, lake water, ground water, and ocean. The effect of human activity on the compositional variations of elements in the hydrosphere and soils will be considered below.

6.2.1 Acid Rain-Soil-Ground Water System

SO_2 and NO_x in atmosphere which are derived by industrial activities dissolve into rainwater to become acid rain (Fig. 6.8). Acid rain falls to the surface, causing for damage of forest (Appendix, Plate 35) and reacts with soils, resulting to the

6.2 Anthropogenic Fluxes to the Hydrosphere and Soils and Mass Transfer Mechanism

dissolution of various elements from soils. pH of rainwater increases through this reaction. Rainwater penetrates downward to become soil water and ground water. Such process will be discussed below.

6.2.1.1 Rainwater-Atmosphere Reaction

CO_2 takes important role to control pH of rainwater. CO_2 in the atmosphere dissolves into rainwater as

$$H_2O + CO_2 = HCO_3^- + H^+ \tag{6.4}$$

Equilibrium constant for this reaction at 25 °C is given by

$$K_{6-4} = m_{H^+} m_{HCO_3^-}/P_{CO_2} = 10^{-7.83} \tag{6.5}$$

where γ_{H^+}, $\gamma_{HCO_3^-}$ and a_{H_2O} are assumed to be unity.

Electroneutrality relation in H_2O–CO_2 system is expressed as

$$m_{H^+} = m_{HCO_3^-} + m_{OH^-} + 2m_{CO_3^{2-}} \tag{6.6}$$

Assuming $\sum CO_2 = m_{HCO_3^-}$, it is approximated that $m_{H^+} = m_{HCO_3^-}$. Giving $P_{CO_2} = 3.1 \times 10^{-4}$ atm (atmospheric P_{CO_2}), and $m_{H^+} m_{HCO_3^-} = 10^{-11.34}$, $m_{H^+2} = 10^{-11.3}$ and pH 5.65.

In recent years, the CO_2 emission due to human activities increased significantly. The change in pH of rainwater caused by human activities is estimated as follows. Above equations give $m_{H^+} = P_{CO_2}^{1/2} \times 10^{-7.83}$. Therefore, pH $= 7.83 \times 1/2 \log P_{CO_2}$. If P_{CO_2} changes to two times of present value, pH is reduced by 0.15.

If P_{CO_2} increases with the rate same to present one, P_{CO_2} becomes two times of present value after 60 years. Therefore, pH does not change significantly only by the effect of increased CO_2. In contrast, the dissolutions of HCl, HNO_3 and H_2SO_4 to rainwater cause considerable decrease in pH of rainwater.

For example, pH of H_2SO_4–H_2O system is determined as follows (Lerman 1979). Dominant dissolved species are H^+, OH^-, HSO_4^-, and SO_4^{2-}. Electroneutrality relation is expressed as

$$m_{H^+} = m_{OH^-} + m_{HSO_4^-} + 2m_{SO_4^{2-}} \tag{6.7}$$

This relation is approximated as

$$m_{H^+} = 2m_{SO_4^{2-}} \tag{6.8}$$

Table 6.5 Kinetic data on reactions in aqueous solution (Earth environmental technology handbook edition committee 1991)

	Reaction	Reaction rate
1.	$H_2O_2 + (SO_2)_{aq} \rightarrow SO_4^{2-} + 2H^+$	$8.0 \times 10^4 \exp[-3{,}650(-1/T - 1/298)]$
2.	$O_3 + S(IV) + H_2O \rightarrow 2H^+ + SO_4^{2-}$	$4.39 \times 10^{11}\exp(-4{,}131/T) + 2.56 \times 10^3\exp(-966/T)/\{H^+\}$
3.	$H_2O \longleftrightarrow H^+ + OH^-$	$1.0 \times 10^{-14}\exp[7{,}153.6(1/298 - 1/T)]$
4.	$CO_2 + H_2O \longleftrightarrow HCO_3^- + H^+$	$4.5 \times 10^{-7}\exp[1{,}544(1/298 - 1/T)]$
5.	$HCO_3^- \longleftrightarrow CO_3^{2-} + H^+$	$4.5 \times 10^{-11}\exp[1{,}744(1/298 - 1/T)]$
6.	$HNO_3 + H_2O \longleftrightarrow NO_3^- + H^+$	$22.0\exp[2{,}371(1/T - 1/298)]$
7.	$HNO_2 \longleftrightarrow NO_2^- + H^+$	5.0×10^{-4}
8.	$NH_3 + H_2O \longleftrightarrow NH_4^+ + OH^-$	17.6×10^{-6}
9.	$SO_2 + H_2O \longleftrightarrow HSO_3^- + H^+$	$1.7 \times 10^{-2}\exp[2{,}037(1/T - 1/298)$
10.	$HSO_3^- \longleftrightarrow SO_3^{2-} + H^+$	$6.3 \times 10^{-8}\exp[1{,}996(1/T - 1/298)]$
11.	$HSO_4^- \longleftrightarrow SO_4^{2-} + H^+$	1.2×10^{-2}
12.	$HCl + H_2O \longleftrightarrow Cl^- + H^+$	1.3×10^6
13.	$Cl + Cl^- \longleftrightarrow Cl_2^-$	2.0×10^5
14.	$(HCOOH)_{aq} \longleftrightarrow HCOO^- + H^+$	1.8×10^{-4}
15.	$(CH_3COOH)_{aq} \longleftrightarrow CH_3COO^- + H^+$	1.73×10^{-5}
16.	$(HCHO)_{aq} + HSO_4^- \longleftrightarrow HMSA^* + H_2O$	8.6×10^{-4}

If $m_{SO_4^{2-}} = 3.6 \times 10^{-6}$ mol·L^{-1}, we obtain $m_{H^+} = 7.2 \times 10^{-6}$ and this means pH 5.1. In the atmosphere in industrial region, SO_4^{2-} concentration of rainwater is high. If $m_{SO_4^{2-}} = 36 \times 10^{-6}$ mol·L^{-1}, we obtain pH 4.1.

pH in HNO$_3$–H$_2$O system is derived to be lower than 5.65 in a manner similar to above case. Most of gaseous species reduce pH. However, the presence of H$_2$S and NH$_3$ causes an increase in pH. In the above cases simple binary system is considered. However, rainwater-atmosphere system is multi-component system.

For H$_2$O–CO$_2$–H$_2$SO$_4$–NH$_3$–HCl system, we should consider the electroneutrality relation as

$$m_{H^+} + m_{NH_4^+} = m_{HCO_3^-} + m_{Cl^-} + m_{NO_3^-} + m_{HSO_4^-} + 2m_{SO_4^{2-}} + 2m_{CO_3^{2-}} + m_{OH^-} \tag{6.9}$$

Fine particles (aerosols such as salts, Kosa etc.) are dispersed in the atmosphere. They dissolve into rainwater. These dissolutions affect pH of rainwater. Dissolved species like Na$^+$, and K$^+$ have to be taken into account for electroneutrality relation and chemical reaction.

Above discussion is mainly based on chemical equilibrium between gaseous species and rainwater. However, kinetics such as dissolution rates of gaseous species into rainwater and of fine particles into rainwater may control chemical compositions and pH of rainwater. A large number of kinetic data on reactions in aqueous solution are available (Table 6.5) and some applications of kinetics have been carried out (e.g., Lerman 1979). However, a few studies on the kinetics of dissolution of fine particle in rainwater has been done.

6.2 Anthropogenic Fluxes to the Hydrosphere and Soils and Mass Transfer Mechanism

Table 6.6 Time to equilibration of a water droplet with a non-reacting gas, time to droplet evaporation, and distance fallen through air (Lerman 1979)

Droplet initial radius r_0 (μm)	Time to steady state for gas diffusion into droplet[b] t_{ss} (s)	Life span in an atmosphere of 80 % relative humidity[a]	
		Time to complete evaporation (s)	Distance fallen (?)
3	0.004	0.16	2 μm
10	0.04	1.8	0.2 mm
30	0.12	16	2.1 cm
100	4.0	290	208 m
150	9.0	900	1.05 km

[a]From Rogers (1976)
[b]Equation (4.37), with $D = 1 \times 10^{-5}$ cm^2 s^{-1}

If equilibrium between atmosphere and water is attained, concentration of dissolved gaseous species is, $m_A = KP_A$, where P_A is partial pressure of gas A and K is constant.

Concentration of gaseous species near atmosphere–water boundary is governed by molecular diffusion rate. Time to reach steady state concentration of gaseous species is expressed as (Table 6.6)

$$t_{ss} = 0.4 r_0^2 / D \tag{6.10}$$

where D is diffusion coefficient (cm^2 s^{-1}) of dissolved gaseous species in water, and r_0 is radius of rainwater drop.

When D is 1×10^{-5} cm^2 s^{-1}, $t_{ss} = 0.4 \times 10^5 r_0^2$. Generally, r_0 is in a range of 10^{-3}–10^{-4} m (Lerman 1979). For example, if r_0 is 150 μm, $t_{ss} = 0.4 r_0^2 / D = 9(s)$. This time is generally shorter than that of rainfall.

In general, the reaction rate between rainwater and gaseous species is fast compared with the rate of evaporation of rainwater. This means, in general, the equilibrium tends to be attained.

6.2.1.2 Rainwater-Soil Reaction

Rainwater evaporates near soil surface. Dissolved HCl and H_2SO_4 do not evaporate and concentrate to water remained. Namely, acidic components enrich near soil surface. Acid solution dissolves alkali (Na, K), alkali earth (Ca, Mg) and Al. With the proceeding of evaporation, some minerals containing these elements (sulfates, carbonates etc.) precipitate near soil surface. If rainwater containing marine chlorides (NaCl, KCl) near the coastal region evaporates, these salts form in the soil. What kind of salts precipitates depends on solubility, concentration in aqueous solution and degree of evaporation.

If soil water evaporates, ground water ascends due to advection-capillary phenomenon. Evaporation of saline ground water leads to accumulation of salts near soil surface. Particularly, near the coastal region, saline ground water ascends, leading to salt accumulation near the surface of soil. If Na^+ and Ca^{2+} ascend to near surface of soil, they are adsorbed by clay minerals to form Na- and Ca-clay minerals. Further supply of salt components causes accumulation of sulfates.

In the arid region where acid rain falls, acid materials (H_2SO_4 etc.) tend to accumulate near soil surface. When rain falls and dissolves acid materials, rainwater and lake water become to be acidic rapidly by the input of acid surface waters. However, in humid region, rapid decrease in pH due to input of acid solution does not occur. In humid region, SO_4^{2-} in rainwater penetrates downwards and becomes to ground water accompanied by its adsorption onto soils and rainwater is neutralized by the ion exchange reaction with soil such as

$$RNa + H^+ = RH + Na^+ \tag{6.11}$$

where RNa is cation exchangeable site.

With proceeding of this ion exchange reaction, buffering capacity for pH reduces. Further proceeding of reaction causes an increase in exchangeable Al^{3+}, resulting to acceleration of Al^{3+} dissolution. Al^{3+} tends to be fixed as low solubility minerals such as gibbsite ($Al(OH)_3$) etc. Al concentration in soil water is determined by the Al crystalline and amorphous phases. In low pH condition, Al concentration in soil water is high that is controlled by high solubility minerals such as Al-sulfates (e.g., alunogen).

Generally, pH of soil water is low due to oxidation of organic matter, producing CO_2 by the reaction

$$CH_2O + O_2 \rightarrow CO_2 + H_2O \tag{6.12}$$

Produced CO_2 dissolves to soil water, resulting to lower pH according to the reaction

$$CO_2 + H_2O \rightarrow H^+ + HCO_3^- \tag{6.13}$$

The reaction (6.12) is caused by bacterial activity. The acid solution in soil dissolves Ca^{2+} from rocks. Increased Ca^{2+} and CO_3^{2-} by these reactions may cause precipitation of $CaCO_3$.

Oxidation of pyrite causes a decrease in pH, that is expressed as

$$FeS_2 + (15/4)O_2 + (7/2)H_2O \rightarrow Fe(OH)_3 + 2SO_4^{2-} + 4H^+ \tag{6.14}$$

Decrease in pH is also caused by the activity of sulfur oxidation bacteria. This reaction is given by

$$H_2S + 2O_2 \rightarrow SO_4^{2-} + 2H^+ \tag{6.15}$$

6.2 Anthropogenic Fluxes to the Hydrosphere and Soils and Mass Transfer Mechanism

pH of acid soil water and ground water increases by the dissolution of silicate ($MSiO_3$; M: divalent cation). This is expressed simply as

$$MSiO_3 + 2H^+ \rightarrow M^{2+} + SiO_2 + H_2O \qquad (6.16)$$

Silicate dissolution occurs also by the following reaction.

$$MSiO_3 + 2CO_2 + H_2O \rightarrow M^{2+} + SiO_2 + 2HCO_3^- \qquad (6.17)$$

In most cases, pH increased in near surface soil environment becomes to neutral-alkaline condition. In addition, the following dissolution of carbonates leads to increasing of pH.

$$CaCO_3 + H^+ \rightarrow Ca^{2+} + HCO_3^- \qquad (6.18)$$

$$CaCO_3 + CO_2 + H_2O \rightarrow Ca^{2+} + 2\,HCO_3^- \qquad (6.19)$$

If ground water whose pH is controlled by the dissolutions of silicates and carbonates, inputs of ground water to river water does not decrease pH of river water. However, if acid surface water inputs to river water, pH of river water reduces and cation concentrations change.

Reductions of anthropogenic SO_4^{2-} to H_2S and NO_3^- to NH_4^+ occur in soils, causing the changes in pH and concentrations of cations in soil water. These reactions occur by oxidation of organic matter and pyrite by bacteria.

Anions combine with cations to form minerals. For example, SO_4^{2-} combines with Ca^{2+} to form gypsum ($CaSO_4 \cdot 2H_2O$). In high pH condition apatite ($CaHPO_4 \cdot 2H_2O$) precipitates by a combination of HPO_4^{2-} with Ca^{2+}. Fluorite (CaF_2) forms by the reaction of F^- with Ca^{2+}. The concentration of Ca^{2+}, HPO_4^{2-}, SO_4^{2-} and F^- in soilwater are sometimes controlled by the solubilities of these minerals (Fig. 6.9).

The change in water quality from acid rain to ground water depends on chemical and mineralogical compositions and physical features of host rocks (e.g., permeability, porosity). For example, dissolution rate of volcanic glass is rapid compared with other silicate minerals. Chemical reaction between volcanic glass and water causes the formation of surface alteration layer composed of amorphous phase containing Si, Al and Fe such as allophane. Cycling of elements in soils relates not only to acid rainwater but also to other various processes. As an example of cycling, S cycling is shown in Fig. 6.10. Influence of microorganisms is particularly important for S cycling. Due to the activity of microorganisms the following changes occur, organic sulfur, SO_4^{2-}, S^{2-}, $S^{2-} \rightarrow S_0$, $S_0 \rightarrow SO_4^{2-}$, SO_4^{2-} $\gamma \rightarrow$ organic S (Table 6.7). Sulfur in soil comes from organic S, SO_4^{2-} in acid rainwater, and SO_4^{2-} and S^{2-} in host rocks. Sulfur cycling is intimately related to the other elemental cycling. For instance, S in pyrite (FeS_2) in host rocks dissolves to aqueous solution as H_2S. H_2S is oxidized by sulfur oxidizing bacteria, leading to decreasing of pH according to the reaction.

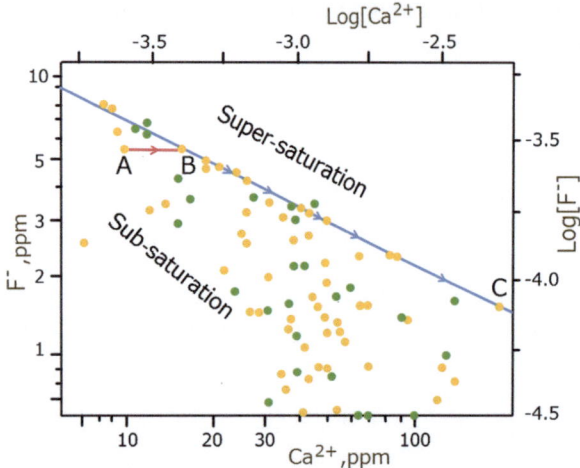

Fig. 6.9 The stability of fluorite and the saturation of ground waters from Sirohi, W. Rajasthan, India. The evolution in water chemistry upon addition of gypsum is described by the pathway A, B to C

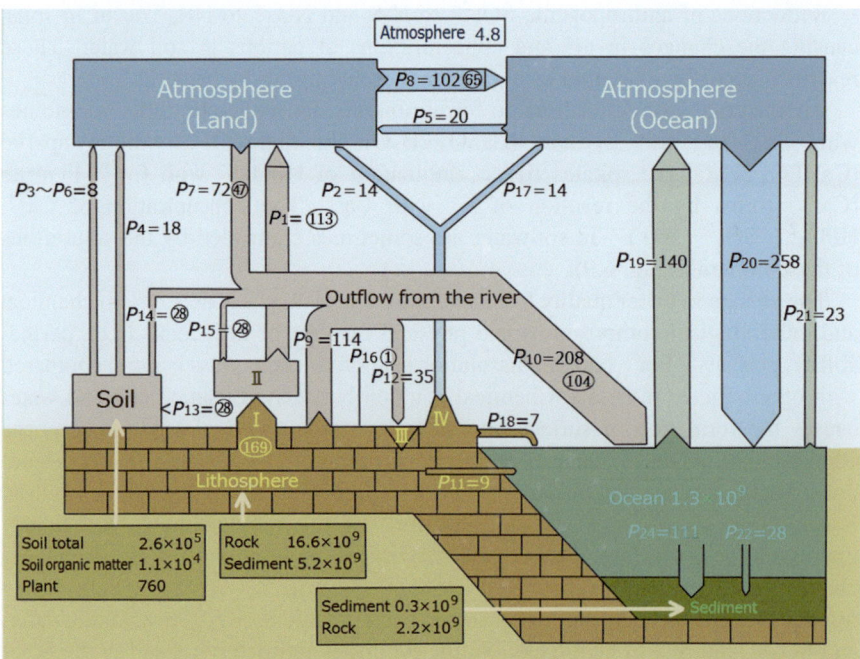

Fig. 6.10 S cycle in soil (Kimura 1989)

6.2 Anthropogenic Fluxes to the Hydrosphere and Soils and Mass Transfer Mechanism

Table 6.7 Microorganisms influencing S cycle (Kimura 1989)

S	Microorganisms
Inorganization of organic S	
(C–O–S) bond $\to$ SO_4^{2-}	Organic nutrition microorganism
(C–S) bond $\to$ SO_4^{2-}	
$S^{2-} \to S^0$	Thiobacillus, Beggiatoa
$S^0 \to S_4^{2-}$	Thiobacillus
$SO_4^{2-} \to S^{2-}$	Desulfovibrio, Desulfotomaculum
$SO_4^{2-} \to$ Organic S	Organic nutrition microorganism (plant)

$$H_2S + 2O_2 \to SO_4^{2-} + 2H^+ \qquad (6.20)$$

Low pH solution formed by this reaction releases elements from silicates.

Base metal element concentration in the atmosphere increases due to human activity. Aerosol containing these base metal elements falls and accumulates on soil surface. The formation of gypsum ($CaSO_4 \cdot 2H_2O$) is accelerated by the catalysis of the base metal elements.

6.2.1.3 Base Metal Concentrations in Ground Water

Anthropogenic organic matter dissolved in ground water consumes O_2 in ground water, causing decreasing of Eh (oxidation-reduction potential). Due to this decreasing, Fe and Mn concentrations in ground water tend to increase. It is obvious in Eh-pH diagram (Figs. 6.11 and 6.12) that this change occurs.

Fe and Mn dissolve from the rocks into ground water in reducing environment. It is inferred that solvation of Fe and Mn by the formation of chelate compounds combined with various organic acids (Matsumoto 1983). High Fe and Mn concentrations of ground water are caused by natural process as well as by anthropogenic process. For example, ground water in coal mining region contains large amounts of Fe and Mn.

Due to the oxidation of ground water with high concentration of Fe and Mn, Fe and Mn hydroxides precipitate. Inorganic oxidation rate for Fe^{2+} and Mn^{2+} is slow, but it is accelerated by bacterial activity. For example, iron oxidizing bacteria takes important role for the reaction (Yamanaka 1992)

$$4Fe^{2+} + O_2 + 4H^+ \to 4Fe^{3+} + 2H_2O \qquad (6.21)$$

The rate of this reaction in acidic condition is higher than inorganic reaction rate but not different in neutral-alkaline conditions (Fig. 6.13). Inorganic and biological oxidation rate is different for Fe and Mn. For instance, if Fe^{2+} and Mn^{2+} oxidation rates in HCO_3^- solution are compared, it is found that Mn^{2+} does not oxidize less than pH 9, while Fe^{2+} oxidizes rapidly less than pH 6.6 (Stumm and Morgan 1970).

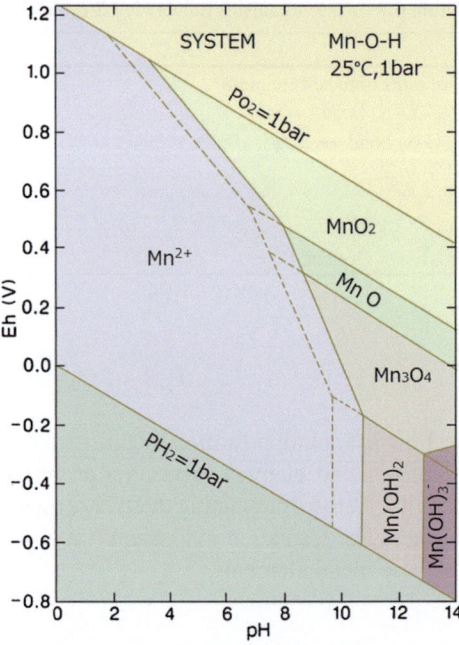

Fig. 6.11 Eh-pH diagram for part of the system Mn–O–H. Assumed activity for dissolved $\sum Mn = 10^{-6}$ (Brookins 1988)

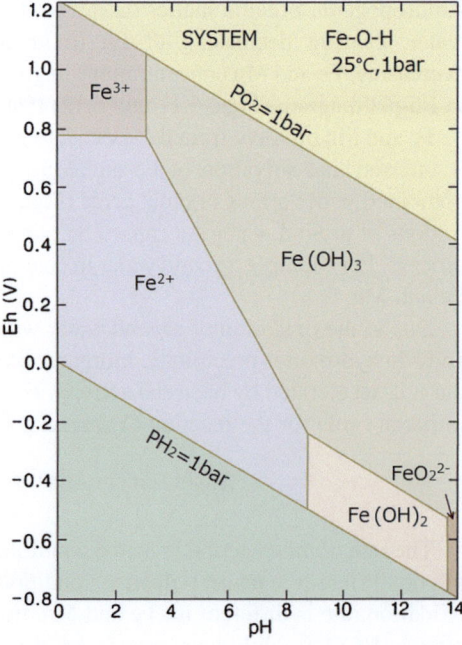

Fig. 6.12 Eh-pH diagram for part of the system Fe–O–H assuming Fe(OH)$_3$ as stable Fe(III) phase. Assumed activity of $\sum$ dissolved Fe $= 10^{-6}$ (Brookins 1988)

Fig. 6.13 Oxygenation rate of ferrous iron as a function of pH (Stumm and Morgan 1970)

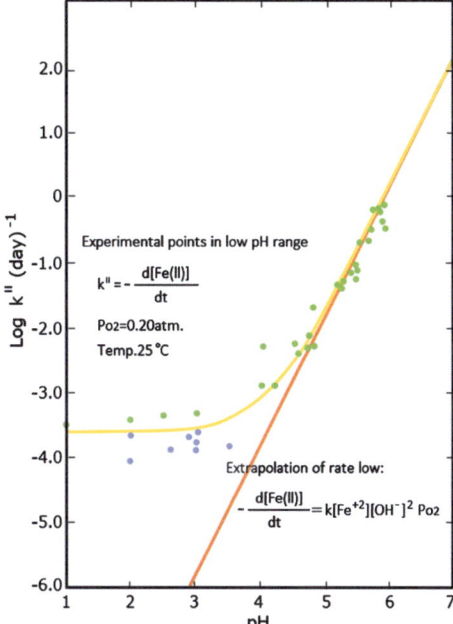

Base metal adsorption capability of Fe- and Mn-hydroxides is high. Fe- and Mn-hydroxides contain sometimes appreciable amounts of other base metals as solid solution (e.g., Cr in Fe $(OH)_3$). Therefore, concentration of base metals in ground water are strongly affected by the presence of Fe- and Mn-hydroxides.

Base metal elements are adsorbed by cation exchangeable sites in clay minerals in soils. The selective adsorption and conservation of base metal ions by cation exchangeable sites occur.

The concentrations of base metal elements in drainage from metal mines are controlled by bacterial activity. For example, dissolution of chalcopyrite and subsequent secondary formation of Cu-sulfates are promoted by iron oxidizing bacteria. This bacteria leaching is used to recover base metals from drainage and ores from mines (Yamanaka 1992).

The behavior of base metal elements is affected by acid rains. As shown in Fig. 6.14, Cd concentration of lake water and soil water is high in low pH lake water (Fujinawa 1991). This high concentration is due to release of Cd adsorbed onto mineral surfaces in acid condition. It is essentially important to elucidate the behavior of base metal elements according to the strong influences of base metal elements in ecological system. Formation of minerals also influence on base metal concentrations in soils. For example, Cd is contained in $CaCO_3$ as a solid solution

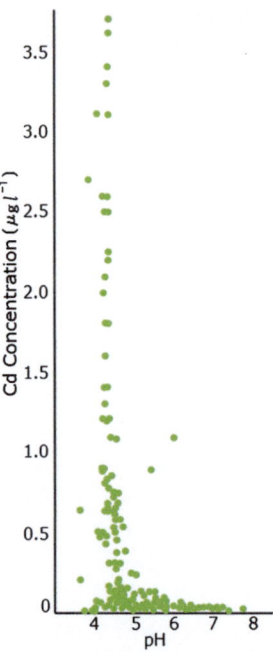

Fig. 6.14 Cd concentration and pH of lake water soil water in Sweden (Fujinawa 1991)

(Appelo and Postma 1993). In this case, Cd concentration of ground water is governed by the reactions

$$CaCO_3 = Ca^{2+} + CO_3^{2-} \tag{6.22}$$

$$CdCO_3 = Cd^{2+} + CO_3^{2-} \tag{6.23}$$

Equilibrium constants for (6.22) and (6.23) are expressed as

$$K_{cc} = a_{Ca^{2+}}\ a_{CO_3^{2-}}/a_{cc} \tag{6.24}$$

$$K_{ota} = a_{Cd^{2+}}\ a_{CO_3^{2-}}/a_{ota} \tag{6.25}$$

where K is equilibrium constant, cc is calcite ($CaCO_3$), ota is otavite ($CdCO_3$), and a is activity.

Combination of (6.24) with (6.25) leads to

$$K_{ota}/K_{cc} = (m_{Ca^{2+}}\lambda_{CaCO_3}X_{CaCO_3})/(m_{Ca^{2+}}\gamma_{Ca^{2+}}\lambda_{CdCO_3}X_{CdCO_3}) \tag{6.26}$$

where γ is ionic activity coefficient, λ is activity coefficient of solid solution component, X is mole fraction of solid solution component, and m is ionic molality. It could be assumed the $\lambda_{Cd^{2+}}$ is equal to $\lambda_{Ca^{2+}}$.

6.2 Anthropogenic Fluxes to the Hydrosphere and Soils and Mass Transfer Mechanism

If regular solution model is applied to calcite-otavite solid solution, λ is expressed as

$$ln\lambda_{CdCO_3} = \alpha_o X^2_{CaCO_3} \qquad (6.27)$$

$$ln\lambda_{CaCO_3} = \alpha_o X^2_{CdCO_3} \qquad (6.28)$$

where α_o is interaction parameter.

Giving $\alpha_o = -0.8$ (Davis et al. 1987; Fuller and Davis 1987) and 1 mol% $CdCO_3$, we obtain

$$ln\lambda_{CdCO_3} = -0.8(0.99)^2 = -0.78, \lambda_{CdCO_3} = 0.46 \qquad (6.29)$$

$$ln\lambda_{CaCO_3} = -0.8(0.01)^2 = -8.0 \cdot 10^{-5}, \lambda_{CaCO_3} = 1.00 \qquad (6.30)$$

Putting $K_{ota} = 10^{-11.31}$, $K_{cc} = 10^{-8.45}$, $m_{Ca^{2+}} = 3 \cdot 10^{-13}$ (6.29), and (6.30) into (6.26), we obtain

$$m_{Cd^{2+}} = (K_{ota}\alpha_{CdCO_3}X_{CdCO_3}m_{Ca^{2+}})/(K_{cc}\alpha_{CaCO_3}X_{CaCO_3})$$
$$= 10^{-11.31} \times 0.46 \times 0.01 \times 3 \cdot 10^{-3}/10^{-8.48} \times 1.00 \times 0.99$$
$$= 2.06 \times 10^{-8} \text{mol L}^{-1} \qquad (6.31)$$

This numerical test (Appelo and Postma 1993) suggests that concentration of Cd^{2+} in ground water is controlled by the formation of $CdCO_3$–$CaCO_3$ solid solution. It is inferred that $CdCO_3$–$CaCO_3$ solid solution forms after the formation of Cd surface complexes (Davis et al. 1987).

6.2.2 Pollution of River Water

Chemical compositions of river water are determined mainly by inputs of ground water and surface water (rainwater, ice melted water), aerosol fall, biological activity, and evaporation. In addition to these processes, anthropogenic influence (Appendix, Plate 36) is increasing in recent years.

Table 6.8 summarize natural and anthropogenic riverine fluxes (Holland and Petersen 1995). Anthropogenic fluxes for many elements exceed natural fluxes. Examples include S, N (ammonia), P, Cu, Zn, Pb, Cr, Sn, Mo, Cd, Hg, Au, and Pt group elements. Most of them are toxic and are derived from metal mines and industries to river water, while anthropogenic effect of Al, Ni and Mg is small. However, it has to be noted that not all of anthropogenic elements move to river water. Concentrations of pollutant in river water are variable in place to place. When waste water from mines comes into river water, and mixes with river water, the precipitation of minerals, adsorption by iron hydroxides and clay minerals and uptake by organisms decrease their concentrations in river water. It is important to

Table 6.8 Natural and anthropogenic riverine fluxes (Holland and Petersen 1995)

Element	1990 World production (gm/year)	Unpolluted river flux (gm/year)	Ratio of 1990 world production to unpolluted dissolved plus particulate river flux
Carbon	$5,000 \times 10^{12}$	400×10^{12} organic 250×10^{12} inorganic	6
Iron	500×10^{12}	$1,000 \times 10^{12}$	0.50
Sulfur	58×10^{12}	100×10^{12}	0.58
Nitrogen (ammonia)	110×10^{12}	10×10^{12}	11
Aluminum	18×10^{12}	$1,600 \times 10^{12}$	0.01
Phosphorus	22×10^{12}	2×10^{12}	11
Copper	9.0×10^{12}	1.1×10^{12}	8.2
Zinc	7.3×10^{12}	1.4×10^{12}	5.2
Lead	3.3×10^{12}	0.26×10^{12}	13
Chromium	11.7×10^{12}	2.0×10^{12}	5.8
Nickel	1.0×10^{12}	6.5×10^{12}	0.2
Tin	220×10^9	60×10^9	3.7
Magnesium	380×10^9	$400,000 \times 10^9$	0.95×10^{-3}
Molybdenum	114×10^9	30×10^9	3.8
Uranium	20×10^9	36×10^9	0.55
Cadmium	21×10^9	4×10^9	5.2
Mercury	6.0×10^9	1.6×10^9	3.7
Gold	2.0×10^9	0.08×10^9	25
Platinum group metals	285×10^6	200×10^6	1.4

Source: World production figures from mineral commodities summaries (1991)

elucidate these removal mechanisms, inorganic and biogenic reactions, and to carry out geochemical modeling, taking into account chemical reaction, fluid flow and diffusion (e.g., Takamatsu et al. 1977). Anthoropogenically polluted and naturally influenced (e.g. weathering) riverwater (Appendix, Plate 37) influence seawater chemistry (see Sect. 4.2).

6.2.3 Pollution of Lake Water

Pollution of lake water is caused by the inputs of polluted river water, ground water and acid water from metal mines. Pollution and acidification of lake water have been considerably investigated by the model analyses. pH and the changes in the water quality of lake water inputted by acid rainwater and acidified river water will be considered below.

6.2.3.1 pH of Lake Water

pH of lake water is controlled by various types of interaction as well as inputs of acid rainwater and acidified river water. They include (1) weathering and acidified

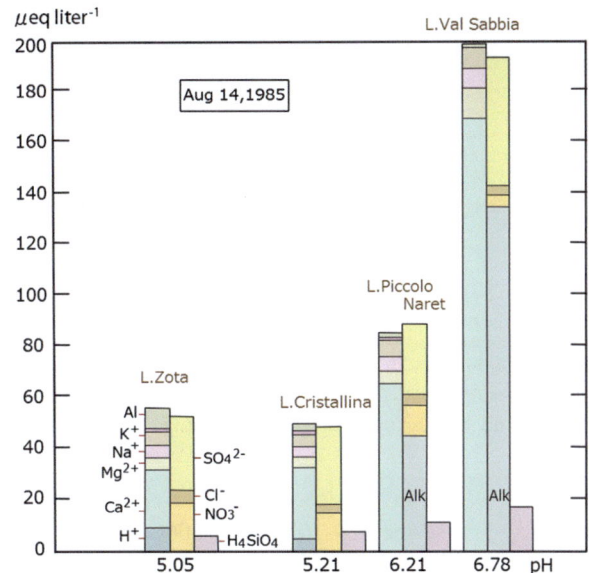

Fig. 6.15 Water composition of four lakes in southern Alps of Switzerland. The difference in composition is caused by the geology of the bedrocks in the catchment areas. Lakes Zota and Cristallina are situated within a drainage area of gneissic rocks. The other two lakes are in catchment areas that contain calcite and dolomite (Stumm and Schnoor 1995)

river water, (1) weathering, (2) ion exchange reaction, (3) oxidation-reduction reaction, and (4) biomass activity (Stumm and Schnoor 1995). pH is enhanced by the weathering reactions such as

$$CaCO_3 \text{ (calcite)} + 2H^+ \rightarrow Ca^{2+} + CO_2 + H_2O \tag{6.32}$$

$$CaAl_2Si_2O_8 \text{ (Ca-feldspar)} + 2H^+ + H_2O$$
$$\rightarrow Ca^{2+} + Al_2Si_2O_5(OH)_4 \text{ (kaolinite)} \tag{6.33}$$

If lake water reacts with lake bottom carbonate (limestone, dolomite), pH increases rapidly because of fast dissolution rates of carbonates (Sect. 6.3.1.1). In contrast dissolutions of silicates (granite, metamorphic rocks etc.) do not proceed considerably because of slow dissolution rate of silicates, and pH of lake water does not change quickly. Figure 6.15 shows that pH of lake water depends on lithology of lake bottom. Examples of ion exchange reactions controlling pH are given by

$$2ROH + SO_4^{2-} \rightarrow R_2SO_4 + 2OH^- \tag{6.34}$$

$$NaR + H^+ \rightarrow HR + Na^+ \tag{6.35}$$

where R is cation exchangeable site.

Due to these reactions, pH value is enhanced. Microorganisms cause oxidation-reduction reactions such as

$$NH_4^+ + 2O_2 \rightarrow NO_3^- + H_2O + 2H^+ \tag{6.36}$$

$$H_2S + 2O_2 \rightarrow SO_4^{2-} + 2H^+ \tag{6.37}$$

$$SO_4^{2-} + 2CH_2O + 2H^+ \rightarrow 2CO_2 + H_2S + 2H_2O \tag{6.38}$$

$$FeS_2 + 15/4O_2 + 7/2H_2O \rightarrow Fe(OH)_3 + 2SO_4^{2-} + 4H^+ \tag{6.39}$$

The reactions (6.36), (6.37), (6.38), and (6.39) lead to the change of pH. The other examples of microorganisms activities controlling pH are reductions of Fe(OH)$_3$ and MnO$_2$, causing increasing pH. These reactions are written as

$$(CH_2O)_{106}(NH_3)_{16}(H_3PO_4) + 424Fe(OH)_3 + 862H^+$$
$$\rightarrow 424Fe^{2+} + 16NH_4^+ + 106CO_2 + HPO_4^{2-} + 1166H_2O \tag{6.40}$$

$$(CH_2O)_{106}(NH_3)_{16}(H_3PO_4) + 212MnO_2 + 398H^+$$
$$\rightarrow 212Mn^{2+} + 16NH_4^+ + 106CO_2 + HPO_4^{2-} + 298H_2O \tag{6.41}$$

As mentioned above, pH changes due to the interactions between water, rocks (minerals), and organisms. Evaporation of lake water in arid region leads to increasing of pH due to a release of gaseous component from lake water, resulting to alkaline lake. Inputs of volcanic gases (SO$_2$, HCl etc.) cause decreasing of pH. In above cases gaseous components control pH of lake water. pH influences solubility of minerals, adsorption and biological activity, resulting to the changes in water quality. Water quality is also related to inputs of river water and rainwater, mixing of fluids, diffusion and fluid flow. Many computer simulation studies taking into account these factors have been carried out and examples treating simple cases will be mentioned below.

6.2.3.2 Perfectly Mixing Non-Steady State Model

Calculation based on perfectly mixing non-steady state model was done by Hanya (1979). Mass balance equation for this model is given by

$$Vdm/dt = q(m_i - m) \tag{6.42}$$

where V is volume (lake water storage amount), m is concentration, q is volume inflow and outflow rate, and m_i is inflow concentration.

6.2 Anthropogenic Fluxes to the Hydrosphere and Soils and Mass Transfer Mechanism 197

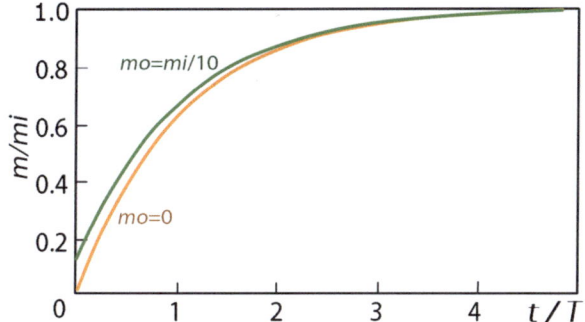

Fig. 6.16 Variation in concentration of lake water due to dilution (perfectly mixing model) (Morita Y; in Hanya T (ed) 1979)

This equation is solved as

$$m = m_0 + \left[\{1 - exp(-t/\tau)\}m_i q/\left(q' - m_0\right)\right] \quad (6.43)$$

where m_0 is initial concentration of lake water prior to inflow of polluted water, and τ is residence time.

Temporal variation in water quality for Eq. (6.43) is shown in Fig. 6.16. This figure illustrates that lake water is significantly polluted after about four times of residence time. Volume inflow rate is same to volume outflow rate in the above case. However, volume inflow rate is generally different from volume outflow rate. In this case

$$V dm/dt = qm_i - q'm \quad (6.44)$$

where q is volume inflow rate, q' is volume outflow rate, and m_i is concentration of input water.

Therefore,

$$m = m_0 + \{1 - exp(-t/\tau)\}\left\{m_i q/\left(q' - m\right)\right\} \quad (6.45)$$

Equations (6.42), (6.43), (6.44), and (6.45) are established for the system which is not affected by precipitation of minerals such as Fe- and Mn-hydroxides, adsorption of dissolved species, colloids and fine grained particles onto minerals, uptake of elements by organisms and dissolution of sediments. Figure 6.17 shows an example of the system which is not affected by these processes. Observed concentration is in agreement with the calculated concentration based on Eq. (6.45) (Fig. 6.17).

Next, we consider the system including more complicated processes including biological activity (Kayane 1972). Water mass balance is expressed as

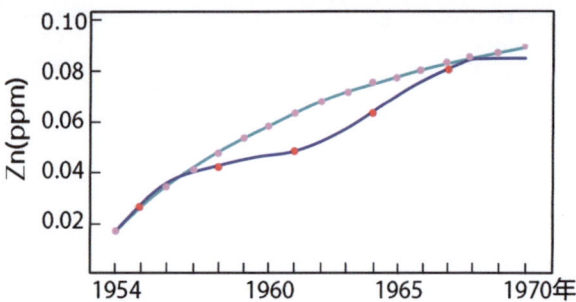

Fig. 6.17 The change in calculated and measured concentration of zinc in Towada lake with time (Morita Y; in Hanya T (ed.) 1979)

$$q_p + q_s + q_g - q_v - q_o = 0 \tag{6.46}$$

where q_p is rainfall to lake surface, q_s is inflow of river water to lake, q_s is ground water inflow to lake, q_v is evaporation of lake water and q_o is outflow of river water from lake. Outflow from lake to ground water is neglected. Mass balance equation for a given element is expressed as

$$m_p q_p + m_s q_s + m_s q_g - m_o q_o + d - s = V dm_i/dt \tag{6.47}$$

where m is concentration, d is transportation rate not by water, and s is removal rate by biological activity, ion exchange and formation of mineral. V is volume of lake water, t is time and m_i is concentration of a given element in lake water.

If lake water is perfectly mixed

$$m_o = m_i \tag{6.48}$$

where m_o is outflow concentration of given element.

Above two equations, (6.47) and (6.48) lead to

$$dm_i/dt + m_i\left\{\left(q_p + q_s + q_g - q_v\right)/V\right\} = \left\{\left(m_p q_p + m_s q_s + m_g q_g + d - s\right)/V\right\} \tag{6.49}$$

From this equation we obtain

$$m_i = \exp\left\{-\int a(t)dt\right\}\int \exp\{a(t)\}b(t)\}dt + c'\exp\left\{-\int a(t)dt\right\} \tag{6.50}$$

where c' is a constant of integration.

Assuming that A and B do not vary with time and $m_I = m_i$ (initial concentration at $t = 0$), we obtain

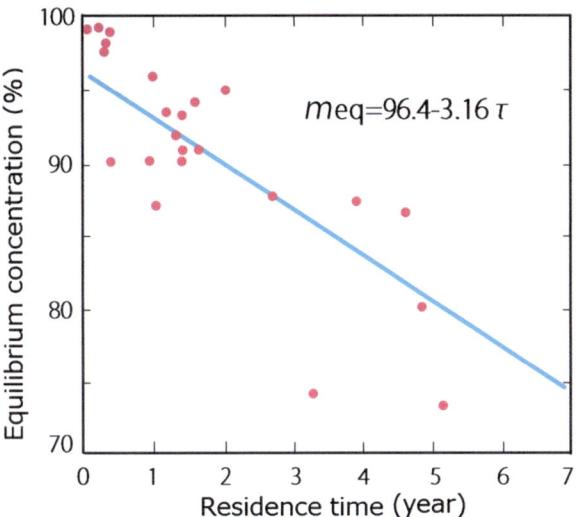

Fig. 6.18 Relationship between residence time and equilibrium concentration of lake water (m_{eq}) in New Hampshire (USA) (Dingman and Johnson 1971; Kayane 1972)

$$m_i = \left(m_p q_p + m_s q_g + m_0 q_0 + d - s\right)/\left(q_p + q_s + q_g - q_v\right)$$
$$+ \left\{m_i - \left(m_p q_p + m_s q_s + m_g q_g + d - s\right)/\left(q_p + q_s + q_g - q_v\right)\right\} \quad (6.51)$$
$$\exp\left\{-\left((q_p + q_s + q_g - q_v)/V\right)t\right\}$$

$$\tau = V/\left(q_p + q_s + q_g - q_v\right) \quad (6.52)$$

where τ is average residence time of perfectly mixing lake water.

Input of pollutant to lake water is assumed to be constant. m_i is expressed as m_{eq} in a case of $t\gamma \to \infty$. In this case, m_{eq} is represented by

$$m_{eq} = \left(m_p q_p + m_s q_s + m_g q_g + d - s\right)/\left(q_p + q_s + q_g - q_v\right) \quad (6.53)$$

Dingman and Johnson (1971) obtained the relation between τ and m_{eq} based on above equations (Fig. 6.18). Figure 6.18 indicates that m_{eq} decreases with increasing τ. q and m can be estimated based on the observation and analyses. However, it is not easy to estimate s because s is a function of several parameters such as rate of precipitations of minerals (e.g., base metal hydroxides).

In the above studies, lake water is regarded as perfectly mixing one box. However, lake water chemistry is heterogeneous. For instance, chemical composition of surface water differs from deep water. Two box (surface water, deep water) model was applied to the interpretation of P concentration in lake water by Brezonik (1994).

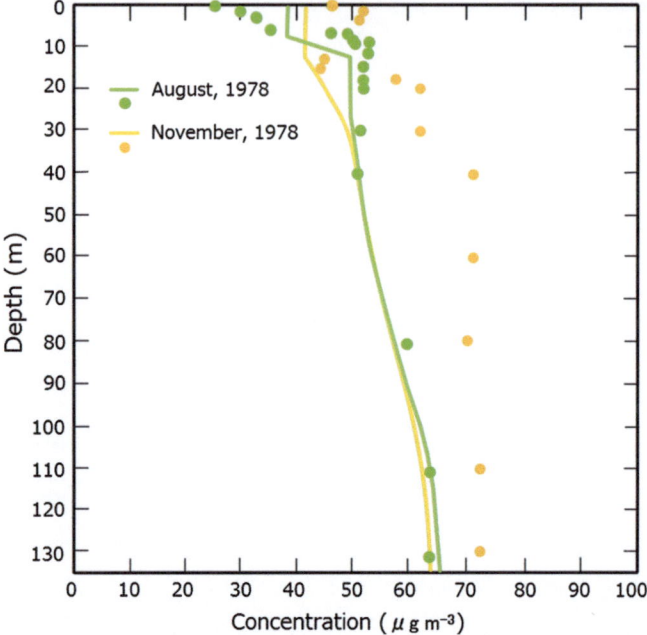

Fig. 6.19 Vertical variation of tetrachloroethylene concentration in Lake Zürich (*Solid line*; calculated values, symbol: measurement values)

6.2.3.3 One Dimensional Vertical Model

Concentrations of minor elements in lake water are usually determined by adsorption onto settling particles, desorption from settling particles and interaction of sediments and bottom water. Thus, one dimensional vertical model assuming laterally homogeneous concentration is useful for the analysis of distribution of minor element concentration. Imboden and Schwarzenbach (1985) calculated tetrachloroethylene concentration in Zurich lake, Switzerland based on this model. Basic equations used by them are mass balance equation concerning solutes and particles, mass balance equation concerning minor elements on particles in sedimentary column, and diffusion equation at the boundary between sediment surface and bottom water. Results of calculation based on these equations and vertical concentration profile are shown in Fig. 6.19.

6.2.4 Pollution in Ocean

Ocean becomes to be polluted due to the input of polluted river water. Pollutants, particularly base metals are enriched in inland sea and closed sea caused by the input of polluted river water (Holland and Petersen 1995). For example, enrichment

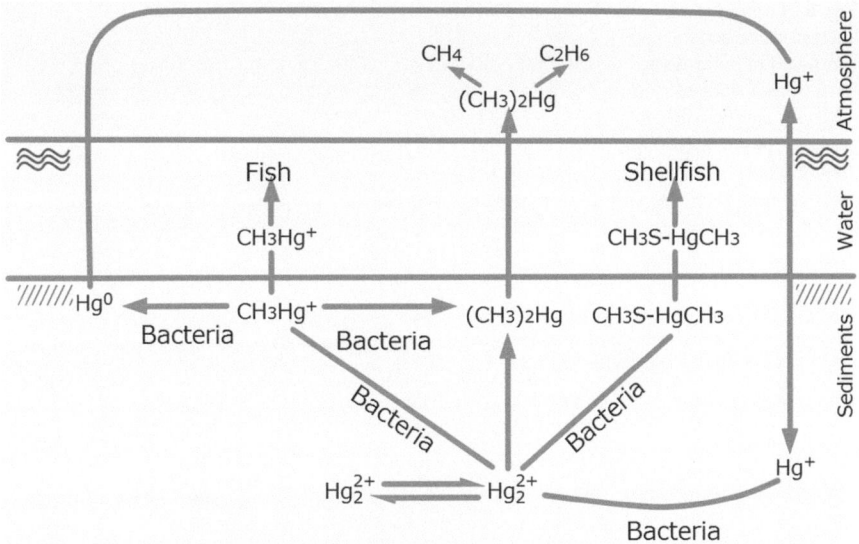

Fig. 6.20 Biocycle of Hg (Wood and Goldberg 1977)

of Cd, Pb, Zn and Cu occurs in Baltic Sea and near coast of U S A. This was caused by disposal of ash generated by burning of coal.

Hg is well studied element due to its toxicity. As shown in Fig. 6.20, chemical state of Hg changes. Role of bacteria is important for this change. Hg in aqueous solution changes to organic Hg(CH_3Hg^+, $(CH_3)_2Hg$, CH_3S–$HgCH_3$) by bacterial activity. The change of organic Hg to inorganic Hg(HgO) also occurs.

Chemical states of base metal elements such as Hg by the interaction of organic matter. For example, methyl B_2 reacts with many base metals, As and Se to form methyl alkyl and enrich to organisms (Wood and Goldberg 1977).

Base metal pollution of ocean is caused not only by the input of polluted river water, but also by dissolution from ship paint. This Cu flux is about 1 % of riverine flux (Goldberg 1976).

6.3 Feedback Associated with Human Waste Emissions

6.3.1 Geological Disposal of High Level Nuclear Waste

Spent fuel from nuclear power stations contains radiogenic U and Pu, and highly radioactive waste solution is produced when the spent fuel is reprocessed. This waste solution needs to be disposed of as it remains highly radioactive for a very long period.

Fig. 6.21 High level nuclear waste body composed of solid-waste form (glass), canister, overpack and backfill and buffer disposed deep underground

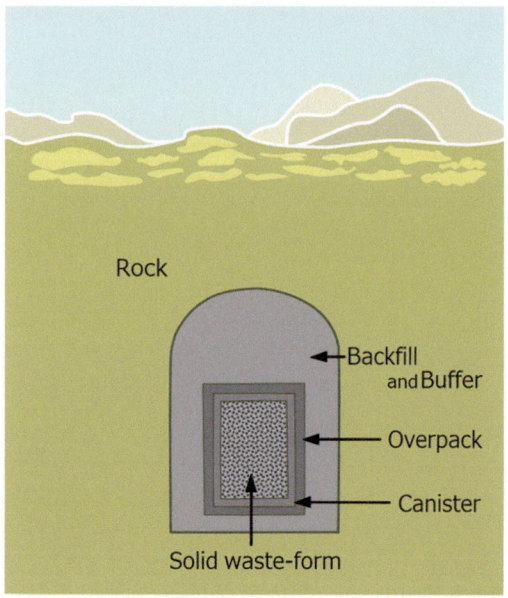

Geological nuclear disposal is considered to be the most favorable disposal option. High-level nuclear waste reprocessed solution is solidified in glass. The waste glass is enclosed in overpack (metals) and buffer material (bentonite) (Fig. 6.21). These are engineered barriers designed to prevent the migration of radioactive elements from the disposal site to the geological environment, although ground water flowing through geological media eventually dissolve these barriers and the glass, and transport the radioactive elements for an extended period (Fig. 6.21). However, the radioactive elements that are transported in this way can be adsorbed by minerals in the rocks, precipitated as minerals, or they can diffuse into micro-fractures (matrix diffusion). Since this adsorption, precipitation, and matrix diffusion all have the effect of retarding the migration of radioactive elements, the migration of radioactive elements in far-field geologic environments is limited.

6.3.1.1 Ground Water Scenario

Groundwater penetrating a repository site will dissolve radioactive elements from the waste glass (Fig. 6.22). The radioactive elements may then be transported for an extended period and over a considerable distance from the repository site by the ground water flow. It is very important to predict both the migration rate, the flux of radioactive elements, and the temporal variation in radioactivity in the near-surface environment.

In order to clarify this process, the chemical composition of the ground water in the buffer material (bentonite), corrosion mechanisms of the overpack, dissolution

6.3 Feedback Associated with Human Waste Emissions

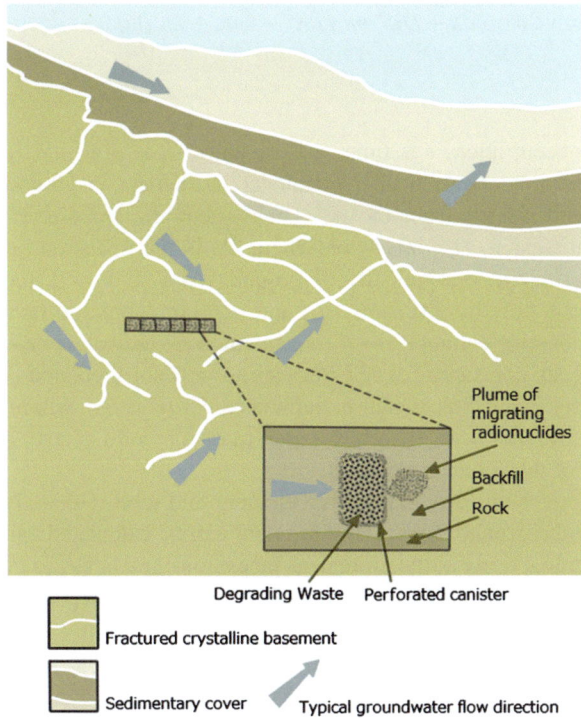

Fig. 6.22 Over a long time period the components of the near-field will degrade and, eventually, the wasteform will be exposed to the ground water and begin to corrode. Dissolution of the wasteform may take place in a locally oxidizing environment due to α-radiolysis of the ground water. Migration of radionuclides may be retarded either by sorption onto the rock, bentonite or the degradation products of the wasteform and canister or by precipitation at the redox front (Miller et al. 1994)

mechanism of glass, and retardation mechanism of radioactive elements in geologic media need to be clarified.

The flux of the radioactive elements due to ground water in each barrier can be estimated using a model designed to consider specifically for each barrier. In this way, the migration behavior of radioactive elements in the buffer material (bentonite) can be inferred using a diffusion model, and the dissolution in glass can be inferred using a solubility model. The solubility model assumes that any radioactive elements moving from the glass into the ground water would precipitate on the surface of the glass, and that dissolved radioactive elements in ground water in contact with the glass are in equilibrium with any radioactive solid compounds on the glass.

The change in the concentration of radioactive elements over time in far-field environments can be expressed as follows (Performance Assessment of Geological Isolation Systems (PAGIS) 1984):

$$\underset{(a)}{\partial\varnothing m_i/\partial t} = \underset{(b)}{v\partial\varnothing m_i/\partial x} + \underset{(c)}{D\partial^2\varnothing m_i/\partial x^2} - \underset{(d_1)}{\lambda_i m_i} + \underset{(d_2)}{\lambda_{i-1}m_{i-1}} - \underset{(e)}{\partial(1-\varnothing)\rho S_i/\partial t} \quad (6.54)$$

where m is concentration, t is time, x is distance, $\varnothing$ is porosity, (a) is temporal variation in the concentration of the radioactive element in ground water, (b) is advection, x is distance from the near field (near the repository site), and (c) is dispersion-diffusion, $D = D_0 + \alpha v$, where D_0 is the diffusion coefficient, α is dispersion coefficient, v is velocity of ground water. If $v = 0$, then radioactive elements migrate only by diffusion. (d_1) and (d_2) correspond to radioactive decay products. Radioactive element, $i - 1$, is generated by radioactive element i, and λ_i and λ_{i-1} are decay constants. (e) describes sorption/desorption, and is related to the distribution coefficient Kd_i, which is defined as $Kd_i = S_i/m_i$, where ρ is apparent density, S_i is the mass of adsorbed i per mass of solid phase, and m_i is the concentration of the solute in solution.

The radioactivity of each radioactive element and total radioactivity as a function of time at the ground surface environment can be calculated using Eq. (6.54).

The total radioactivity at the ground surface environment has to be below safety levels. Retardation of the migration rate of the radioactive element is needs to be considered in order to ensure that radioactivity levels remain within safety limits. Therefore, the selection of appropriate disposal sites and engineered barriers (buffer material, overpack, and glass) is very important.

If we consider the Peclet number, which can be defined as $Pe = v L/(D_{eff} + D)$, where v is the true fluid velocity, L is an arbitrarily selected characteristic length, D_{eff} is the effective diffusion coefficient, and D is the dispersion coefficient, it is generally considered that mass transport is affected by diffusion and not by advection. The equation for the mass transport in bentonite is expressed as

$$R_i\partial m_i/\partial t + \partial m_{pi}/\partial t = D_{pi}\partial^2 m_i/\partial r^2 + (1/r)\partial m_i/\partial r - \lambda_i R_i m_i \\ + \lambda_{i-1}R_{i-1}m_{i-1} - \lambda_i m_{pi} + \lambda_{i-1}m_{pi-1} \quad (6.55)$$

where m_i is the concentration of radioactive element i, m_{pi} is the concentration of radioactive element i in pore water, D_{pi} is the diffusion coefficient of radioactive element i in pore water, λ_i is the decay constant of radioactive element i, R_i is the retardation coefficient of radioactive element i, t is time, r is the distance from the center of the waste glass, and R_i is expressed as

$$R_i = 1 + (1-\varepsilon)\rho Kd_i/\varnothing \quad (6.56)$$

where $\varnothing$ is the porosity of buffer material, ρ is the density of buffer material, and Kd_i is the distribution coefficient of radioactive element i.

Using Eqs. (6.55), (6.56) and the initial and boundary conditions, the migration of radioactive elements in ground water in near- and far-fields, as well as the temporal variations radioactivity near the biosphere and humans, can be calculated.

However, the calculation procedure described in (6.55) and (6.56) has several associated problems.

(1) Dissolution, precipitation and sorption equilibria are typically not attained, which means that dissolution and precipitation kinetics of radioactive nuclides need to be considered.
(2) Uncertainties associated with the parameter values that are used in Eqs. (6.55) and (6.56) need to be carefully evaluated, and accurate data have to be obtained. For example, relatively little solubility and rate information is currently available for reduced environments.
(3) Validation of analytical results is necessary. Experimental studies on the migration of radioactive elements from an underground storage facility need to be conducted and findings need to be compared with equivalent scenarios in natural water environments and computer simulations. However, relatively few such studies have been conducted to date. While simple modeling analyses have been conducted on natural water–rock systems, numerous unresolved problems related to migration of radioactive elements in more complicated natural systems remain.
(4) The assumptions above are based on the assumption that the geological environment is stable for extended periods. However, geologic events such as volcanic eruptions, faulting, climate change, weathering, uplift, erosion, earthquakes, and sea level changes occur. Evaluation of the long-term stability of the geological environment is therefore important. The flow of ground water changes over time depending on the cracks formed by earthquakes and faulting activity. Ground water migrating through fractures interacts with surrounding rocks forming alteration minerals that then fill existing fractures. By filling the cracks in this way, these alteration minerals decrease the flow rates of ground water. Changes in the configuration and fractal dimensions of cracks, mineral precipitation rates, and ground water flow rates all contribute to changing the diffusion rate of radionuclides through the rock.
(5) Processes that have not been extensively studied include radionuclide transport by ion partitioning (e.g. precipitation of solids in solution, surface ion exchange), and colloidal and biological reactions.

6.3.1.2 Natural Analogue Studies

Evaluation of long-term safety assessments based on model predictions is difficult. While it is possible to evaluate the accuracy of a model by comparing the results with experiments on radionuclide migration behavior in underground facilities, such underground experiments are typically short-term (up to several years) investigations.

On the other hand, natural analogue studies are useful for clarifying the long-term migration behavior of radionuclides. Such studies include long-term mass transfer from natural geological bodies, such as uranium deposits, volcanic glass, archeological materials or bentonite deposits, which are analogous to the wastes

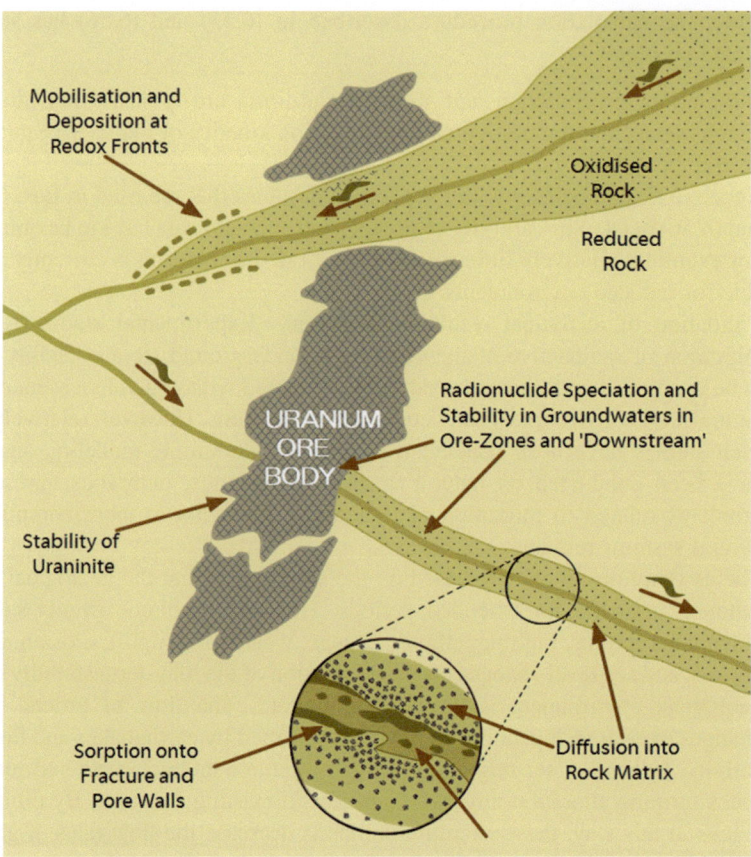

Fig. 6.23 Diagrammatic view of a generic uranium ore body showing the principal processes of interest in natural analogue studies in this geological environment (Miller et al. 1994)

materials employed at underground storage facilities (glass, metal, buffer material). Natural analogue studies can also be used to examine differences at varying temporal and spatial scales. For example, a natural analogue study treating the radionuclides migration from waste body is the study on migration of elements such as uranium from uranium deposits to geologic environment. The migration of radionuclides from an underground storage facility could be compared to the migration of uranium from uranium deposits in a natural geological setting (e.g., Laverov et al. 2009) (Fig. 6.23). An example of this study was conducted at the Oklo uranium deposits in Gabon (Brookins 1984). The Oklo uranium deposits occur in Precambrian rocks where their formation at 2 Ga was associated with the formation of a natural nuclear reactor. Brookins (1978) investigated the mobilization of elements formed by nuclear fission in this natural nuclear reactor and found that most of elements (actinides, Pb, Bi etc.) have not mobilized very far away from the uranium deposits, despite the long duration after the formation of the uranium deposits and high temperatures in the area.

Other studies have examined the effect of igneous intrusive rocks on the distribution of elements in host rocks. For example, Brookins et al. (1982) reported that there was no noticeable mobilization of elements analogous to radionuclides under conditions of high temperature and oxygen isotopic exchange between intrusive rocks and host rocks. In a study on granitic rocks intercalated with mineral layers and their interactions with hydrothermal solution, it was found that rare earth elements (REE) and uranium are fixed in alteration minerals in granitic rocks, which implies that granitic bodies are effective natural barriers to radionuclide migration.

Although these studies examined the long-term behavior (10^5–10^6 years) of elements and isotopes qualitatively, more quantitative, short-term studies ($<10^5$ years) on natural analogues are considered necessary. For example, Yusa et al. (1991) and Shikazono and Takino (2002) investigated the chemical interactions between ground water and natural basaltic glass; a reaction that is considered to be a natural analogue of high-level nuclear-waste glass.

Other natural analogues have been reported in environments such as clay deposits, alkali mineral springs, and archeological materials.

Studies on the geochemical migration behavior of natural analogue elements (e.g., light rare earth elements (LREE) analogous to Am and Cm) are also well suited to studies on predicting the long-term behavior of radioactive elements released from the glass. Shikazono et al. (2003) and Shikazono and Ogawa (2009) analyzed the release of REE from weathered rocks (weathered granite, volcanic soils) and found that, while significant amounts of LREE are released during chemical weathering, the elements are adsorbed and/or become bound to secondary minerals (e.g. clay minerals, iron hydroxide) by ion exchange. Their findings suggested that Am and Cm, which are chemically analogous to LREE, as well as LREE, do not migrate far away from the waste disposal site over the long term. Previous studies determined K_d and the retardation factor R associated with the adsorption of radionuclides on the minerals.

Over the long term, coprecipitation of trace elements onto growing crystals in far-field geological settings near nuclear waste storage facilities is considered to play an important role in limiting the migration of radionuclides, such as Am, Cm and U, that are released from nuclear waste glass and spent fuel. Secondary minerals, such as carbonates (e.g., calcite), sulfates (e.g., gypsum) and clay minerals (e.g., smectite), all tend to incorporate radionuclides by ion partitioning. For instance, Dobashi and Shikazono (2008) analyzed calcite in sedimentary rocks from the Tono uranium mine area in Japan for REE and U. Compared with bulk host rocks, they found that the LREE and U were significantly enriched in calcite, indicating that the incorporation of Am, Cm and U into calcite could likely retard the long-term migration of these elements. More detailed discussion on the role of solid solution formation in relation to the long-term safety of nuclear waste disposal can be found in Bosbach (2010).

In addition to the far- and near-field studies on natural analogues mentioned above, the studies on radionuclide migration mechanisms (e.g., solubility, speciation, sorption, adsorption, ion exchange, precipitation, matrix diffusion, colloid, redox front, biological activity, gas generation) are also regarded as the natural analogue studies.

6.3.2 Underground CO_2 Sequestration

It is generally considered that CO_2 injection into aquifers is a potentially viable method for reducing greenhouse gases (e.g., Gunter et al. 1997). If CO_2 is injected into deep aquifers (ca. 1,000 m depth), it becomes CO_2 supercritical fluid (CSF). CSF changes to CO_2 gas after coming into contact with ground water (Spycher and Reed 1989), and this CO_2 gas and CSF are trapped in the pores of rocks (residual gas trapping). CO_2 dissolves into H_2O liquid (ground water) according to the following reaction:

$$CO_2 + H_2O \rightarrow H_2CO_2 \qquad (6.57)$$

$$H_2CO_3 \rightarrow HCO_3^- + H^+ \qquad (6.58)$$

This process is called solubility trapping, and the concentration of dissolved HCO_3^- depends on pH and P_{CO_2}. pH increases when ground water interacts with silicates, as follows:

$$CaSiO_3 \text{ (wollastonite)} + H_2O + 2H^+ \rightarrow Ca^{2+} + H_4SiO_4 \qquad (6.59)$$

$$CaSiO_3 + 2H_2CO_3 + H_2O \rightarrow Ca^{2+} + 2HCO_3^- + H_4SiO_4 \qquad (6.60)$$

$$MgSiO_3 \text{ (Mg-pyroxene)} + 2H_2CO_3 + H_2O$$
$$\rightarrow Mg^{2+} + 2HCO_3^- + H_4SiO_4 \qquad (6.61)$$

$$Mg_2SiO_4 \text{ (Mg-olivine)} + 2H_2CO_3 \rightarrow 2Mg^{2+} + 4HCO_3^- + H_4SiO_4 \qquad (6.62)$$

$$NaAlSi_3O_8 \text{ (Na-feldspar)} + H_2CO_3 + 5H_2O$$
$$\rightarrow Na^+ + 1/2Al_2Si_2O_5(OH)_4 + 2H_4SiO_4 \qquad (6.63)$$

$$CaAl_2Si_2O_8 \text{ (Ca-feldspar)} + 2H_2CO_3 + H_2O$$
$$\rightarrow Ca^{2+} + Al_2Si_2O_5(OH)_4 + 2HCO_3^- \qquad (6.64)$$

In the dissolution of carbonates, CO_2 changes to HCO_3^- according to the reaction:

$$CaCO_3 + H_2CO_3 \rightarrow Ca^{2+} + 2HCO_3^- \qquad (6.65)$$

6.3 Feedback Associated with Human Waste Emissions

$$CaMg(CO_3)_2 + 2H_2CO_3 \rightarrow Ca^{2+} + Mg^{2+} + 4HCO_3^- \qquad (6.66)$$

$$FeCO_3 + HCO_2 \rightarrow Fe^{2+} + 2HCO_3^- \qquad (6.67)$$

If the ground water becomes over-saturated with carbonates, then dissolution results in the precipitation of the carbonates according to the following reactions:

$$\left.\begin{array}{l} Ca^{2+} + 2\,HCO_3^- \rightarrow CaCO_3 + CO_2 + H_2O \\ Ca^{2+} + Mg^{2+} + 4\,HCO_3^- \rightarrow CaMg(CO_3)_2 + CO_2 + 2H_2O \\ Fe^{2+} + 2\,HCO_3^- \rightarrow FeCO_3 + CO_2 + H_2O \end{array}\right\} \qquad (6.68)$$

Thus, through this mineral trapping, CO_2 can be trapped almost permanently underground.

In addition, the common carbonate mineral dawsonite ($NaAlCO_3(OH)_2$) commonly forms under these conditions (Xu et al. 2005), although the formation thereof is probably short-lived (Hellevang et al. 2005).

Geochemical modeling methods have been used to estimate the amount of CO_2 that can be stored in an aquifer by solubility and mineral trapping. Most of the geochemical modeling simulations have been conducted based on the following equation (e.g., Perkins and Gunter 1995; Talman et al. 2000)

$$R = R_{acid} + R_{neutral} + R_{base}$$
$$= A^* \left(k_a^+ (H^+)^x + k_n^+ k_b^+ (OH^-)^y (P_{CO_2})^2 (1 - \Omega^n)^m \right) \qquad (6.69)$$

where R is the dissolution rate of the mineral, R_{acid}, $R_{neutral}$ and R_{base} are the dissolution rates under acidic, neutral and basic conditions, respectively, Ω is the saturation index, A^* is the surface area of mineral/kg · H_2O, x, y and 2 are reaction coefficients that have been empirically determined, n and m indicate the reaction order, and k_{a^+}, k_{n^+} and k_{b^+} are reaction rate constants for reactions under acidic, neutral and alkaline conditions, respectively.

The reaction rate constants ($k_{a^+}, k_{n^+}, k_{b^+}$) are the most important parameters affecting solubility and mineral trapping. Consequently, numerous experiments have been conducted to examine the reaction rates (dissolution) for different rock types (granite, volcanic rocks (basalt etc.), sedimentary rocks etc.) and minerals as a function of pH, temperature etc. (e.g., Shikazono et al. 2012; Adachi and Shikazono 2009; Umemura et al. 2014a, b).

The geochemical computer models, EQ3/6 (Wolery 1978), SOLMINEQ.88 (Kharaka et al. 1988), PATHARC.94 (Perkins and Gunter 1995) and TOUGHREACT (Xu et al. 2000, 2004) have been used to model water–rock reactions driven by the formation of carbonic acid when waste CO_2 is injected into deep aquifers using experimentally determined rate data (e.g., Gunter and Perkins 1993). These calculations have shown that carbonate aquifers are limited in the quantity of CO_2 that can be trapped, while siliciclastic aquifers have superior potential for trapping CO_2 through the precipitation of carbonates, particularly, when they contain basic aluminosilicates (feldspars, zeolites, illites, chlorites and smectites).

PATHARC.94 (Perkins and Gunter 1995) has been used to interpret and extrapolate the findings of laboratory experiments on CO_2 mineral trapping and bicarbonate brine formation over longer time periods (Gunter et al. 1997). The initial information required to run PATHARC.94 includes the formation water chemistry, the mass of each of the minerals in equilibrium with formation water, and the mass, grain size, and dissolution rate constants for each reactant phase. For these calculations, the following total rate of reaction (Lasaga 1981, 1994) was used (Gunter et al. 1997):

$$R = k(A/M)(1 - \Omega) \qquad (6.70)$$

where Ω is the saturation quotient, k is the dissolution rate constant, A is surface area and M is mass of water.

Perkins and Gunter (1995) reported that these CO_2-trapping reactions would take 100 year to complete after the formation water has equilibrated at the temperature of the glauconitic sandstone aquifer (i.e. 54 °C) in the Alberta Basin (Canada) and at the proposed injection pressure for the CO_2 (260 bars).

Although numerous geochemical modeling simulations have been conducted for underground CO_2 sequestration (see Marini 2007), the results have a large associated uncertainty, primarily due to the uncertainties associated with parameters affecting solubility and mineral trapping (e.g., rate constant (k), surface area (A), mass of water (M), precipitation rate (nucleation, crystal growth etc.). For example, the dissolution rate constant for basalt, which is considered to be one of the most suitable host rock types for underground CO_2 sequestration (Shikazono et al. 2011) in a closed system is significantly lower than that in an open (flow-through) system (Shikazono et al. 2012; Umemura et al. 2014a, b).

Cited Literature

Adachi Y, Shikazono N (2009) Shigenchishitsu 59:9–21 (in Japanese with English abstract)
Ahn J (1988) Mass transfer and transport of radionuclides in fractured porous rock. D Sc Univ California, Berkley
Bosbach D (2010) Solid-solution formation and the long-term safety of nuclear-waste disposal. In: Prieto M, Stoll H (eds) EMU notes in mineralogy, vol 10, pp 325–344
Brookins DG (1978) Chem Geol 23:309–323
Brookins DG (1990) Gabon Waste Manag 10:285–296
Brookins DG, Abashian MS, Cohen LH, Wolenberg HA (1982) In: Topp SV (ed) Scientific basis for nuclear waste management, vol V. North-Holland Press, Amsterdam
Butterfield DA, Massote GJ, McDuff RE, Lupton JE, Lilley MD (1990) J Geophys Res 95:12894–12921
Christensen UR, Yuen DA (1984) J Geophys Res 89:4389–4402
Croviser JL, Honnorez J, Eberhart JP (1987) Geochim Cosmochim Acta 51:2987–2990
Davies CW, Jones AL (1955) Trans Farady Soc 57:872–877

Davis JA, Fuller CC, Cook AD (1987) Geochim Cosmochim Acta 51:1437–1490
Dingman SC, Johnson AH (1971) Wat Res 7:1208–1215
Dobashi R, Shikazono N (2008) Chikyukagaku (Geochem) 42:79–98 (in Japanese with English abstract)
Dykhuizen RC, Casey WH (1989) Geochim Cosmochim Acta 53:2797–2805
Feiss PG (1978) Econ Geol 73:397–404
Field RJ, Noyas RM (1977) Acc Chem Res 10:214–221
Fleming BA (1986) J Colloid Interface Sci 110:40–64
Flicher M, Ross J (1974) J Chem Phys 60:3458–3465
Fuller CC, Davis JA (1987) Geochim Cosmochim Acta 51:1491–1502
Garrels RM, Christ CL (1965) Solutions minerals and equilibria, A Harper International Student Reprint, Harper & Roe, John Weatherhill. Thermodynamics
Goldberg ED (1976) The health of the oceans. The UNESCO Press, Paris
Gunter WD, Perkins EH (1993) Energy Convers Manag 34:941–948
Gunter WD, Wiwchar B, Perkins EH (1997) Miner Petrol 59:121–140
Han MW, Suess E (1989) Palaeogeogr Palaeoclimatol Palaeoecol 71:97–118
He S, Oddo JE, Tomson MB (1994a) J Colloid Interface Sci 162:297–303
He S, Oddo JE, Tomson MB (1994b) J Colloid Interface Sci 163:372–378
Hellevang H, Aagaard P, Oelkers EH, Kvamme B (2005) Environ Sci Technol 39:8281–8287
Hitchon B (1966) Aquifer disposal of CO_2. Geoscience Publishing, Alberta, Canada
Hofmann AW (1972) Am J Sci 272:69–90
Hosking KFG (1951) Trans R Geol Soc Cornwall 18:309–356
Husar RB, Husar JD (1985) J Geophys Res 90:1115–1125
Imboden DM, Gaechter R (1978) Ecol Model 4:77–98
Imboden DM, Schwarzenbach RP (1985) In: Stumm W (ed) Chemical processes in lakes. Wiley, New York, pp 1–29
Joesten R (1977) Geochim Cosmochim Acta 41:649–670
Kashiwagi H, Shikazono N (2005) Palaeogeograph Palaeoclim Palaeoecol 199:167–185
Kharaka YK, Gunter WD, Agaarwal PK, Perkins EH, DeBraal JD (1988) SOLMINEQ.88: a computer program for geochemical modeling of water–rock interactions. U S Geological Survey. Water Resources Invest-Rep, pp 88–4227
Kimura M (1989) In: Nihonkagakukai (ed) Chemistry of soil, pp 129–146 (in Japanese)
Lasaga AC (1981) Rate laws of chemical reactions. In: Lasaga AC, Kirkpatrick RJ (eds) Kinetics of geochemical processes, Rev Mineral 8. Mineralogical Society of America, Washington DC, pp 1–67
Laverov NP, Velichkin VI, Fujiwara A, Shikazono N, Aloshin AP, Asadulin EE, Golubev VN, Kryloba TL, Pek AA, Chernyshev IV (2009) Res Geol 59:342–358
Leeman WP, Carr MJ, Morris JD (1994) Geochim Cosmochim Acta 58:149–168
Mason RP, Fitzgerald WF, Morel FMM (1994) Geochim Cosmochim Acta 58:3191–3198
Matsumoto E (1983) Chikyukagaku (Geochem) 17:27–32 (in Japanese with English abstract)
Morris J, Leeman WP, Tera F (1990) Nature 344:31–36
Muehlenbachs K (1977) Can J Earth Sci 14:771–776
Muehlenbachs K, Clayton RN (1976) J Geophys Res 81:4365–4369
Nordstrom DK (1989) Geochim Cosmochim Acta 53:1727–1740
Perkins EH, Gunter WD (1995) Aquifer disposal of CO_2-rock greenhouse gases: modeling of water-rock reaction paths in a siliciclastic aquifer. In: Perkins EHG, Gunter WD, Hitchon B, Solmineq GW (eds) Introduction to ground water geochemistry. Geoscience Publishing Ltd, Alberta, Canada
Plummer LN, Wiglley TML, Parkhurst DL (1979) In: Jenne EA (ed) Chemical modeling in aqueous systems. A. C. S. Smp. Ser. No.93. American Chemical Society, Washington DC, Chap. 25, pp 539–573
Seyfried WE Jr (1987) Annu Rev Earth Planet Sci 15:317–335
Shikazono N (2002) J Geogr (Chigaku Zasshi) 111:55–65 (in Japanese with English abstract)

Shikazono N, Ogawa Y (2009) Genshiryoku Backend Res 12:3–9 (in Japanese with English abstract)
Shikazono N, Takino A (2002) Res Back End Nucl Energy 8:171–178 (in Japanese with English abstract)
Shikazono N, Ohtani H, Kimura S (2003) Shigenchishitsu 53:201–206 (in Japanese with English abstract)
Shikazono N, Harada Y, Kashiwagi H, Ikeda N (2012) Basalt: types, petrology & uses. Nova Publisher, New York
Shikazono N, Umemura T, Arakawa T (2013) Goldschmidt conference
Silver PG, Carlson RW (1988) Annu Rev Earth Planet Sci 16:477–541
Simmons SF, Christensen BW (1994) Am J Sci 294:361–400
Sohnel O, Mullin JW (1988) J Colloid Interface Sci 123:43–50
Southern JR, Hay WW (1977) J Geophys Res 82:3825–3842
Spycher NF, Reed MH (1989) Econ Geol 84:328–359
Sterrn KH (1954) Chem Rev 54:79–99
Stumm W, Schnoor JL (1995) In: Lerman A, Imboden D, Gat J (eds) Physics and chemistry of lakes. Springer, Berlin, pp 185–215
Stumm W, Furrer G, Kanz B (1983) Croat Chem Acta 56:593–611
Sverjensky DA (1984) Earth Planet Sci Lett 67:70–78
Tanaka T (1994) In: Energy and Environment Research Group (ed) Consideration on CO_2 problem. Nihonkogyo Shinbunsya, pp 129–140 (in Japanese)
Tarney J, Pickering KT, Knipe RJ, Dewey JD (eds) (1991) Philosophical transactions of the royal society London A, vol 335, pp 227–418
Umemura et al. (2014a) Appl Geochem (submitted)
Umemura et al. (2014b) Geochem J (submitted)
Uyeda S, Kanamori H (1979) J Geophys Rev 84:1049–1061
Villas RN, Norton D (1977) Econ Geol 72:1471–1504
Von Damm KL, Edmond JM, Measure CI, Walden B (1985a) Geochim Cosmochim Acta 49:2197–2220
Von Damm KL, Edmond JM, Measure CI, Grant B (1985b) Geochim Cosmochim Acta 49:2221–2237
Walker JCG (1985) Orig Life 16:117–127
Walsh PR, Duce RA, Fashing JL (1979) J Geophys Res 84:1719–1726
Walshe MP, Bragart SL, Schechter RS, Lake LW (1984) Am Inst Chem Eng J 30:317–328
Walther JV, Wood BJ (1984) Contrib Min Petrol 88:246–259
Weare JH, Stephens JR, Eugster HP (1976) Am J Sci 276:767–816
Welch SA, Ullman WJ (1996) Geochim Cosmochim Acta 60:2939–2948
Wells JT, Ghiorso MS (1991) Geochim Cosmochim Acta 55:2467–2481
Weres O, Yee A, Tsao L (1981) J Colloid Interface Sci 84:379–402
Weres O, Yee A, Tsao L (1982) Society of petroleum engineers J February 9–16
White DE, Muffler LJP, Truesdell AH (1971) Econ Geol 66:75–97
Wolery TJ (1978) Some chemical aspects of hydrothermal processes at mid-oceanic ridges—a theoretical study. Ph.D. thesis Northwestern University
Woley TJ (1979) Calculation of chemical equilibrium between aqueous solutions and minerals: the EQ3/6 software package. Report UCRL-52658. Lawrence Livermore National Laboratory, Livermore
Wood JM, Goldberg ED (1977) In: Stumm W (ed) Global chemical cycles and their alterations by man. Springer-Verlag, Berlin, pp 137–153
Xu T, Apps JA, Pruess K (2000) Lawrence Berkeley National Laboratory Report LBNC-47315, Berkeley
Xu T, Apps JA, Pruess K (2004) Appl Geochem 19:917–936
Yusa Y, Arai T, Kamei G, Takano H (1991) J Atomic Ener Soc Jpn 33:890–905

Further Reading

Appelo CAJ, Postma D (1993) Geochemistry, ground water and pollution. A. A. Balkema, Rotterdam/Brookfield

Brezonik PC (1994) Chemical kinetics and process dynamics in aquatic system. CRC Press, Boca Raton

Brookins DG (1988) Eh-pH diagrams for geochemistry. Springer-Verlag, New York

Brookins DG (1984) Basis of nuclear waste disposal, geochemical approach. Gendaikogakusya (in Japanese) (trans Ishihara T, Ohashi H)

Brookins DG, Abashian MS, Cohen LH, Wollenberg HA (1982) In: Topp SV (ed) Scientific basis for nuclear waste management, vol V. North-Holland Press, Amsterdam

Bunce NJ (1991) Environmental chemistry, 2nd edn. Wuerz Publishing Ltd., Winnipeg, Canada

Fujinawa K (1991) Polluted ground water. Kyoritsu Press (in Japanese)

Garrels RM, Mackenzie FT, Hunt C (1975) Chemical cycles and the global environment – assessing human influences. William Kaufman, Los Alamos, Cal., pp 1–206

Hanya T (ed) (1979) Mechanism of pollution of water. Kyoritsu Press (in Japanese)

Holland HD (1978) The chemistry of the atmosphere and oceans. Wiley, New York/Chichester/Brisbane/Toronto

Holland HD (1984) Chemical evolution of the atmosphere and oceans. Princeton University Press, Princeton

Holland HD, Petersen U (1995) Living dangerously. Princeton University Press, Princeton

Imboden DM, Schwarzenbach RP (1985) In: Stumm W (ed) Chemical processes in lakes. Wiley, New York, pp 1–29

IPCC (1994) In: Houghton JT et al (eds) Climate change 1994: radioactive forcing of climate change and an evaluattion of the IPCCIS92 emission scenarios. Cambridge University Press, Cambridge

Kayane I (1972) Circulation of water. Kyoritsu Press (in Japanese)

Kimura M (1989) In: Chemical Society of Japan (ed) Chemistry of soils, pp 126–146 (in Japanese)

Lasaga AC (1997) Kinetic theory in the earth science. Princeton University Press, Princeton

Lerman A (1979) Geochemical processes—water and sediment environments. Wiley, New York

Lerman A, Imboden D, Gal J (1995) Physics and chemistry of lakes, 2nd edn. Springer-Verlag, Berlin

Marini L (2007) Geological sequestration of carbon dioxide. Elsevier, New York

Miller W, Alexander R, Chapman N, Mckinley I, Smellie J (1994) Natural analogue studies in the geological disposal of radioactive wastes. Elsevier, Amsterdam

Nishimura M (1991) Environmental chemistry. Syokabo, Tokyo (in Japanese)

Nordstrom DK, Munoz JL (1985) Geochemical thermodynamics. The Benjamin/Cummings Publishing Co, Menlo Park

P. A. G. I. S. (1984) Summary report of phase 1, a common methodological approach based on European data and models. VI. EUR 9220

Perkins EH, Gunter WD (1995) A user's manual for PATHARC. 94; a reaction path-mass transfer program. Alberta Research Council Report ENVTRC. 95–11, Wiley, New York, p 179

Stumm W (1977) Global chemical cycles and their alterations by man. Dahlem Konferenzen, Berlin

Stumm W (ed) (1985) Chemical processes in lakes. Wiley, New York

Stumm W (ed) (1987) Aquatic surface chemistry. Wiley, New York

Stumm W, Morgan JJ (1970) Aquatic chemistry. Wiley, New York

Takamatsu T, Naito M, Fan L-T (1977) Environmental system technology. Nikkan Kogyo Shinbunsya, Tokyo (in Japanese)

Yamanaka T (1992) Introduction to biogeochemistry. Gakkaishuppan Center, Tokyo (in Japanese)

Afterwords

In this book we considered mass transfer and elemental migration between the atmosphere, hydrosphere, soils, rocks, biosphere and humans in earth's surface environment on the basis of earth system sciences. In Chaps. 2, 3, and 4, fundamental theories (thermodynamics, kinetics, coupling model such as dissolution kinetics-fluid flow modeling, etc.) of mass transfer mechanisms (dissolution, precipitation, diffusion, fluid flow) in water-rock interaction of elements in chemical weathering, formation of hydrothermal ore deposits, hydrothermal alteration, formation of ground water quality, seawater chemistry. However, more complicated geochemical models (multi-components, multi-phases coupled reaction-fluid flow-diffusion model) and phenomenon (autocatalysis, chemical oscillation, etc.) are not considered.

In Chap. 5, fundamental equations on geochemical cycle and examples of global carbon and other (S, Sr) cycles and climate change deduced from the global carbon cycle simulations are presented. However, recent investigations of coupled geochemical cycles (C, S, Sr, O etc) are not described.

In Chap. 6, the influence of anthropogenic activities on the flux to the atmosphere, hydrosphere and soils and associated mass transfer mechanism are discussed. Geological disposal of high level nuclear disposal and CO_2 underground sequestration are briefly presented.

Finally, for readers who want to study and understand more deeply earth and planetary system science and each field of earth and planetary sciences which are described in this book, I would like to recommend the following outstanding books.

Kinetics

Lasaga AL (1997) Kinetic theory in the earth sciences. Princeton University Press, Princeton

Environmental Geochemistry, Geochemistry

Chapman NA, McKinley IG, Hill MD (1987) The geological disposal of nuclear waste. Wiley, Chichester
Holland HD, Trekian KK (eds) (2004) Treatise on geochemistry. Elsevier, Amsterdam
Holland HD, Turekian KK (eds) (2004) The teatise in geochemistry. Elsevier, Amsterdam
Langmuir D (1997) Aqueous environmental geochemistry. Prentice Hall, Englewood Cliffs
Mason B (1958) Principles of geochemistry, 2nd edn. Wiley, Chichester
Shikazono N (1997) Chemistry of earth system. University of Tokyo Press, Tokyo (in Japanese)
Shikazono N (2010) Environmental geochemistry of earth system. University of Tokyo Press, Tokyo (in Japanese)
Stumm W, Morgan JJ (1996) Aquatic chemistry, 3rd edn. Wiley, Chichester
Zhu C, Anderson G (2002) Environmental applications of geochemical modeling. Cambridge University Press, Cambridge

Economic Geology

Barnes HL (ed) (1997) Geochemistry of hydrothermal ore deposits. Wiley, Chichester
Shikazono N (2003) Geochemical and tectonic evolution of arc-backarc hydrothermal systems. Implication for the origin of kuroko and epithermal vein-type mineralization and the global geochemical cycle. Developments in Geochemistry, vol 8. Elsevier, Amsterdam

Groundwater Geochemistry

Appelo CAJ, Postma D (1993) Geochemistry, ground water and pollution. A. A. Balkema, Rotterdam/Brookfield
Drever JI (1988) The geochemistry of natural waters, 2nd edn. Prentice Hall, Englewood Cliffs

Earth and Planetary System Science and the Global Geochemical Cycle

Berner RA (2004) The phanerozoic carbon cycle. Oxford University Press, New York
Chameides WC, Perdue EM (1997) Biogeochemical cycles. Oxford University Press, New York
Ernst WG (ed) (2000) Earth systems. Cambridge University Press, Cambridge
Kump LR, Kasting JR, Crane RG (1999) The earth system. Pearson-Prentice Hall, Upper Saddle River
NASA (1986) Earth system science: overview—a program for global change. Earth System Science Committee, NASA Advisory Council, Washington DC
Shikazono N (1992) Introduction to earth system. University of Tokyo Press, Tokyo (in Japanese)
Shikazono N (2009) Introduction to earth and planetary system science. University of Tokyo Press, Tokyo (in Japanese)
Shikazono N (2012) Introduction to earth and planetary system science—new view of earth, planets and humans. Springer, Berlin

Earth's Environment and Resources

Craig JR, Vaughan DJ, Skinner BJ (1988) Resource of the earth. Prentice Hall, Englewood Cliffs
Holland HD, Petersen U (1995) Living dangerously. Princeton University Press, Princeton
Marini L (2007) Geological sequestration of carbon dioxide. Elsevier, Amsterdam
Shikazono N (2003) Geochemical and tectonic evolution of arc-back arc hydrothermal system: implication for the origin of Kuroko and epithermal vein-type mineralizations and the global geochemical cycle. Developments in Geochemistry, vol 8. Elsevier, Amsterdam
Skinner BJ (1976) Earth resources, 2nd edn. Prentice Hall, Englewood Cliffs

Hydrosphere

Berner EK, Berner RA (1978) The global water cycle: Geochemistry and environment. Prentice Hall, Englewood Cliffs
Holland HD (1978) The chemistry of the atmosphere and oceans. Wiley, Chichester
Millero FJ (1996) Chemical oceanography, 2nd edn. CRC, Boca Raton
Shikazono N (2012) Science of water resources. Ohmusha (in Japanese)

Biosphere and Soils

Birkeland PW (1999) Soils and geomorphology, 2nd edn. Oxford University Press, New York
Bolt GH, Blluggenwent MG (eds) (1978) Soil chemistry. Elsevier, Amsterdam
Schesinger WH (1997) Biogeochemistry. An analysis of global change. Academic, Waltham
Vernadsky VI (1997) The biosphere. Copernicus Books, New York

Appendix (Plate)

Plate 1 Feldspar Crystal: © 2009 OrthoclaseBresil by Didier Descouens. Feldspar is the most abundant rock-forming mineral, occurring in wide varieties of rocks (igneous, sedimentary and metamorphic rocks). Feldspar is Na-K-Ca-Al silicate. Common feldspar is albite ($NaAlSi_3O_8$), orthoclase ($KAlSi_3O_8$) and anorthite ($CaAl_2Si_2O_8$) (see Chap. 1)

Plate 2 Quartz Crystal: © 2010 Quartz Brési by Didier Descouens. Quartz is the most abundant rock-forming mineral (SiO_2) as well as feldspar, occurring in wide varieties of rocks (igneous, sedimentary and metamorphic rocks) and ore deposits (see Chaps. 1 and 2)

Plate 3 Calcite Crystal: © 2013 Double Refraction in Calcite by N. Takeuchi. Calcite is the most common Ca-carbonate ($CaCO_3$), occurring in wide varieties of geologic environment (limestone, carbonatite, sedimentary rocks, hydrothermal ore deposits) (see Chap. 1)

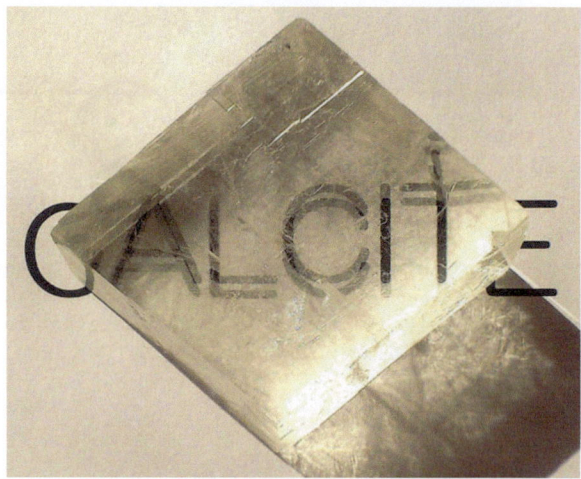

Plate 4 Rhodochrosite Crystal: © 2006 The Searchlight Rhodochrosite Crystal by Eric Hunt. Rhodochrosite is manganese carbonate ($MnCO_3$), occurring in hydrothermal ore deposits (vein-type deposits) (see Chap. 1)

Plate 5 Soil with 4 ft. Depth from the Surface. Soil is a geologic body that is primarily composed of minerals in parent rock, mixed with organic matter. Usually the upper horizon is more intensely weathered than lower horizon (see Chaps. 1 and 6)

Plate 6 Olivine Crystal: © 2010 Forsterite-223825 by Rob Lavinsky. Olivine is Mg-Fe silicate ($(Mg, Fe)_2 SiO_4$). It is common rock-forming mineral, occurring in basic and ultramafic rocks (e.g. basalt, peridotite) (see Chap. 1)

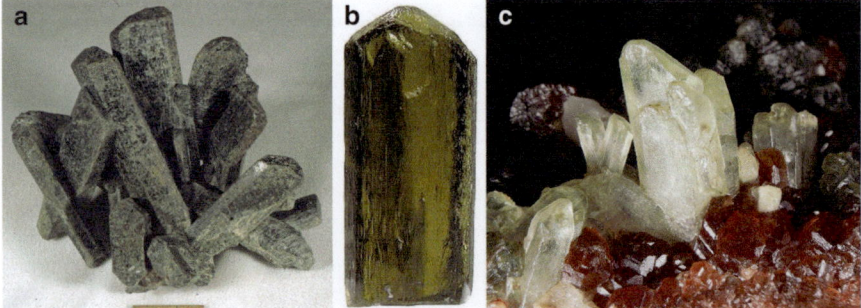

Plate 7 Pyroxene Crystal. 7a: © 2010 Augite-La Panchita Mine, Mun de La Pe, Oaxaca, Mexico by Rock Currier. 7b: © 2010 Enstatite-83152 by Rob Lavinsky. 7c: © 2011 Diopside Aoste by Didier Descouens. Pyroxene is Ca-Mg-Fe silicate ($(Ca, Mg, Fe) SiO_3$) and classified as augite ($(Ca, Mg, Fe) Si_2O_6$ (Plate 2A), enstatite ($MgSiO_3$) (Plate 2B), diopside ($CaMgSi_2O_6$, (Plate 2C). It occurs in volcanic rocks (basalt, andesite) and skarn-type deposits (see Chap. 1)

Appendix (Plate) 223

Plate 8 Columnar jointed Basalt and Waterfall (Svartifoss, Iceland): © 1996 SvartifossSummer by Andreas Tille. Svartifoss (Black Fall) is a waterfall Vatnajokull National Prack in Iceland. Basalt is the most common igneous rocks. The joint is perpendicular to the surface of the flow and is formed by the cooling of the flow (see Chap. 1)

Plate 9 Kakita River, located in Shimazu Town, Shizuoka, central Japan. The river water is originated from rain and snow fall onto Mt. Fuji which is the highest mountain in Japan (see Chap. 1)

Plate 10 Karst Landscape in Minerve, Herault, France: © 2005 Karst minerve by Hugo Soria. Karst topography is a geologic formation caused by the dissolution of carbonate rocks (limestone or dolomite). Weathering resistant rocks such as quartzite also forms Karst topography (see Chap. 1)

Plate 11 Gypsum Crystal: © 2010 Gypsum-209991 by Rob Lavinsky. Gypsum is the most abundant sulfate ($CaSO_4 \cdot 2H_2O$), occurring in evaporite and hydrothermal ore deposits (e.g. Kuroko-type deposits). Gypsum forms by the evaporation of seawater, soilwater and mixing of hydrothermal solution with seawater (see Chap. 2)

Appendix (Plate)

Plate 12 Chalcopyrite:
© 2010 Chalcopyrite
1 by Lloyd.james0615.
Chalcopyrite is cupper-iron
sulfide ($CuFeS_2$), occurring
in hydrothermal ore
deposits (e.g. submarine
hydrothermal ore deposits,
Kuroko-type deposits)
(see Chap. 2)

Plate 13 Pyrite Crystal:
© 2009 2780M-pyrite1
by CarlesMillan. Pyrite is
the most abundant sulfide
(FeS_2), occurring in various
geologic environment
(e.g. hydrothermal ore
deposits, marine sediments,
sedimentary rocks) (see
Chaps. 2 and 4)

Plate 14 Yellow Ore from Matsuki Kuroko Deposits, Akita, northern Japan. About 20 cm in length (Shikazono 1988; Photo by M. Shimizu). The ore is mainly composed of pyrite and chalcopyrite (see Chap. 2)

Plate 15 Sphalerite Crystal: © 2010 Sphalerite-221270 by Rob Lavinsky. Sphalerite is zinc sulfide (ZnS), occurring in hydrothermal ore deposits (e.g. submarine hydrothermal ore deposits, Kuroko-type deposits) (see Chap. 2)

Appendix (Plate)

Plate 16 Galena Crystal. Galena is lead sulfide (PbS), occurring in hydrothermal ore deposits (e.g. submarine hydrothermal ore deposits (e.g. submarine hydrothermal ore deposits, Kuroko-type deposits) (see Chap. 2)

Plate 17 Barite Crystal: © 2011 6158M-barite2 by Carlesmillan. Barite is one of the common sulfate ($BaSO_4$), occurring in hydrothermal ore deposits (submarine hydrothermal ore deposits, Kuroko-type deposits, Mississippi-type deposits) (see Chap. 2)

Plate 18 Gold (*yellow*) with Quartz (*white*): © 2007 Naturkundemuseum Berlin-Gediegen Gold in Quarz, Eagles Nest Mine, Placer County, Kalifornien, USA by Raimond Spekking

Plate 19 Au–Ag Ore from Chitose Au–Ag Vein-Type Deposits, Hokkaido, northern Japan (Shikazono 1988; Photo by M. Shimizu). About 15 cm in length. *Yellowish*, *black-gray* and *white band* is mainly composed of Au–Ag minerals (electrum, argentite), Ag·Pb·Zn sulfides (argentite, galena, sphalerite) and quartz, respectively. Gold grade of this ore is highly enriched more than 1 kg/ton (see Chap. 2)

Plate 20 Photomicrograph showing occurrence of electrum (Courtesy of Sumitomo Metal Mining Co.) (Shikazono et al. 1993). The sample was collected from the Gingro band, Hosen No. 1 vein, 70mLW10. *El* electrum, *Cp* chalcopyrite, *Nm* naumanite (see Chap. 2)

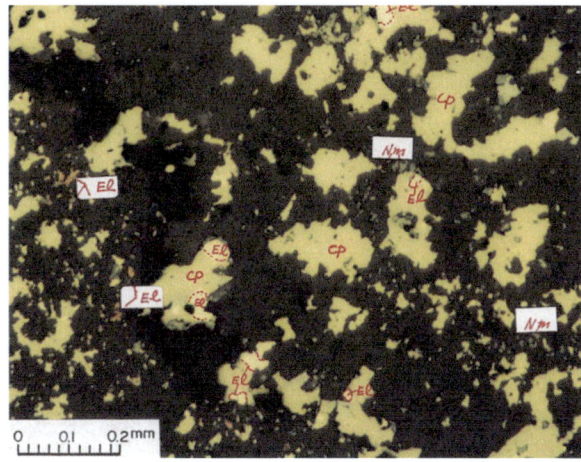

Plate 21 Quartz-adularia-clay vein containing gold (Shikazono et al. 1993). Location: Hishikari, Honzan deposit, Hosen No. 1 vein, 25 mL, 2 Slice W17BW, Vein width: 2.37 m, Ore grade: 1,250 Au g/t, 465 Ag g/t, Wall rock: Sandstone (Shimanto group), *Vein* High grade ore of the Hishikari champion vein, banded quartz vein associated with abundant clay minerals. Electrum occurs in quartz vein and clay band (see Chap. 2)

Plate 22 Arsenic Sulfide Forming near the Lake Shore of Usori in Osorezan Volcanic Area, Aomori, northern Japan (Shikazono et al. 1993; Photo by M. Aoki) (see Chap. 2)

Plate 23 The Osorezan-Type Gold Deposit (Shikazono et al. 1993; Photo by M. Aoki). Hydrothermal activity in the Osorezan caldera, northern Japan, represents a unique example of volcano hosted gold mineralization. Several hot springs in the area precipitate arsenic sulfide which contains significant concentration of gold, showing that the hydrothermal system is an active analog of epithermal gold deposits (see Chap. 2)

Plate 24 Banded Pb–Zn Ore from Hosokura Vein-Type Deposits, Miyagi, northern Japan (Shikazono 1988; Photo by M. Shimizu). About 15 cm in length (see Chap. 2)

Plate 25 Hydrothermal Vent (Black Smoker in Atlantic Ocean) (see Chap. 4). Hydrothermal solution mixes with ambient cold seawater, resulting to the precipitation of fine-grained sulfides black in color

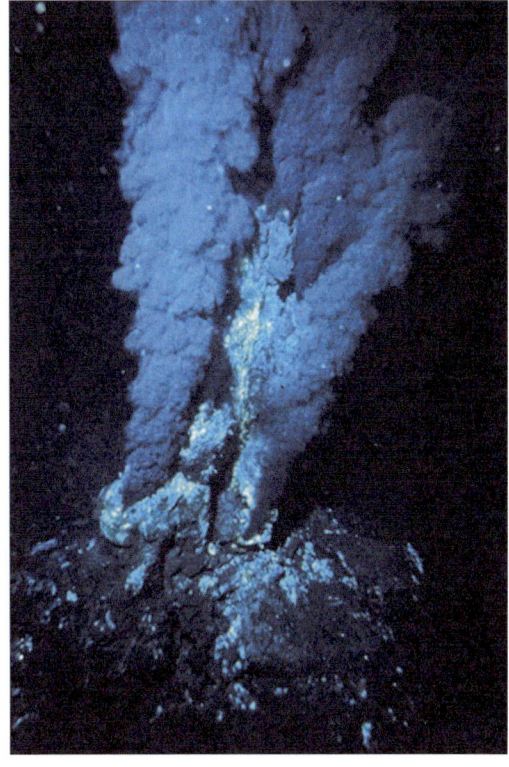

Plate 26 Coral Reafs: © 2010 Coral Outcrop Flynn Reef by Toby Hudson. Coral reafs are community of living organisms and are the most diverse system (see Chap. 5)

Plate 27 Planktonic foraminifera (see Chap. 5)

Appendix (Plate) 233

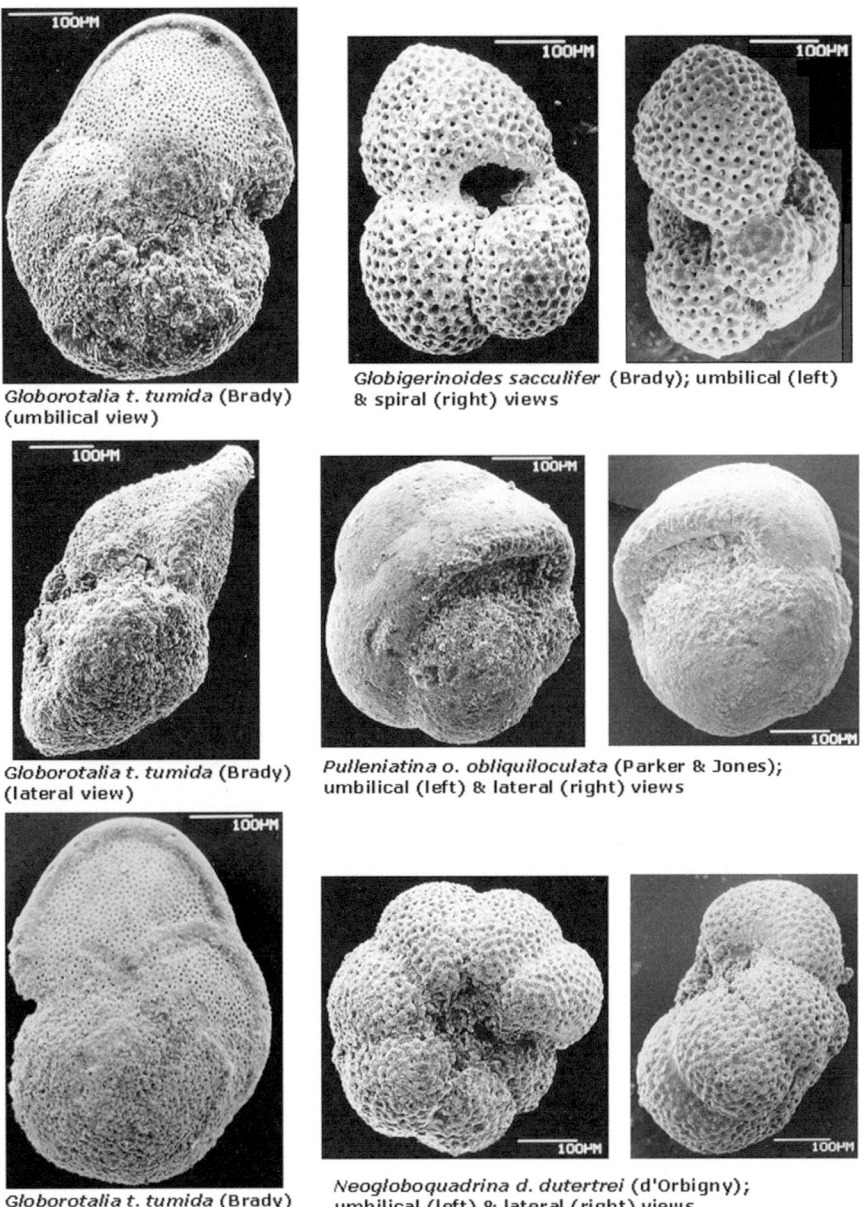

Globorotalia t. tumida (Brady)
(umbilical view)

Globigerinoides sacculifer (Brady); umbilical (left)
& spiral (right) views

Globorotalia t. tumida (Brady)
(lateral view)

Pulleniatina o. obliquiloculata (Parker & Jones);
umbilical (left) & lateral (right) views

Globorotalia t. tumida (Brady)
(spiral view)

Neogloboquadrina d. dutertrei (d'Orbigny);
umbilical (left) & lateral (right) views

Plate 28 Planktonic foraminifera (micrograph) (see Chap. 5)

Plate 29 Mount Fuji and Cherry Blossoms: © 2010 Lake Kawaguchiko Sakura Mount Fuji 4 by Midori. Mt. Fuji is located in Yamanashi and Shizuoka Prefecture, central Japan and is composed of basaltic materials (lava, ash and pyroclastics) (see Chap. 5)

Plate 30 Volcanic Eruption of Augustine Volcano, Alaska. Augustine lies within the area of uplift resulting from the 1964 Alaska earthquake. The volcano is a lava dome complex on Augustine Island in southern western Cuok Inlet in the Kenai Peninsula Borohgh of south central coastal Alaska (see Chap. 5)

Plate 31 Strombolian Volcanic Eruption: © 1980 Stromboli Eruption by Wolfgang Beyer. Strombolian consists of ejection of incandescent cinder lapilli and lava bombs. Stromboli is located in the Aeolian Island of Italy (see Chap. 5)

Plate 32 Castle Geyser at Yellowstone, National Park, Wyoming, United States: © 2008 Steam Phase eruption of Castle geyser with double rainbow by Brocken Inaglory. More than 1,000 geysers have erupted in Yellowstone National Park. Geysers are generally located in volcanic area and associated with silica sinter precipitated from erupting water (see Chap. 5)

Plate 33 Upper Yosemite Fall and Granite at Yosemite National Park, in Yosemite Valley, a National Heritage Site: © 2008 Yosemite falls smt by Smtunli, Svein-Magne Tunli. From the top of the upper fall to the base of the lower fall is 2,425 ft., Wyoming, U.S.A. It is the highest waterfall in north America. The exposed geology of the Yosemite area is dominantly granitic rocks with some older metamorphic rocks (see Chap. 5)

Plate 34 Air Pollution by Industrial Waste Gas (see Chap. 6)

Appendix (Plate) 237

Plate 35 Damage of Forest by Acid Rain at Waldschaeden, Germany: ©1998 Waldschaeden Erzgebirge by bdk (see Chap. 6)

Plate 36 River water pollution (see Chap. 5)

Plate 37 Yellow River Delta. Huang He (Yellow River) in China is the most sediment-filled river. Millions of tons of the soil are carried away by the river every year (see Chap. 6)

Index

A
Acid rain, 173, 174, 182, 186, 187, 191, 194
Actinolite, 53
Activated site, 74
Activation energy, 75–77
Active geothermal system, 23, 67
Activity, 1, 3, 4, 9, 13, 18–20, 28, 30, 37, 39, 44, 58, 66, 76, 77, 95, 98, 99, 106, 107, 109, 116, 121, 123, 130–132, 136, 142, 147–150, 157, 159, 163, 166, 173, 174, 177, 178, 182, 186, 189, 191, 193, 196–198, 201, 202, 204, 205, 207, 208
Activity coefficient, 1, 20, 44, 192
Adiabatic ascending, 56, 105
Adiabatic expansion, 108
Adsorption, 15, 73, 79, 80, 82, 84, 191, 193, 197, 200, 202, 207, 208
Advanced argillic alteration, 23
Advection, 73, 75, 87, 133, 186, 204
Aerosol fall, 28, 122, 123, 127
Aerosols, 184
Allophane, 25, 187
Alteration minerals, 21, 205
Alteration zonings, 24
Aluminum, 17, 24
Alunite, 25
Amorphous $Fe(OH)_3$, 8, 10
Amorphous silica, 21, 22, 83, 110, 115, 117
Amphiboles, 17
Anhydrite, 19, 22, 49, 54, 56, 63, 64, 66, 105, 106, 110, 113, 116, 118, 128, 134, 136, 151
Anthropogenic riverine fluxes, 193
Argillic alteration, 23, 24
Arsenic (As) flux, 165, 166

Atmosphere, 30, 127, 141, 143–147, 152, 156–158, 160, 163, 166, 173, 174, 176–185
Average world river water, 123
Avogadro number, 47

B
Back arc basins, 96, 132, 134, 136, 166–168
Barite, 49, 59, 64, 79, 80, 110–114, 118, 128, 151, 168
Barium (Ba), 167–168
Base metal concentrations in ground water, 189–193
Base metal elements, 103, 105, 128, 136, 166, 189, 191, 201
Batch model, 87
Batch system, 8
Bentonite, 202–204
Biomass activity, 195
Biosphere, 141, 143–145, 176, 204
Black smoker, 113, 114
BLAG model, 147–151, 163
Boiling, 21, 38, 57–61, 69, 70, 108
Bornite, 38, 40, 41
Boron (B) flux, 167
Box model, 145
Buffering capacity, 186
Buffer material, 202–204, 206

C
Ca flux, 136
Calcite, 12–14, 19, 21–23, 45, 46, 57, 58, 80–82, 119, 123, 126, 131, 147, 148, 160, 193, 195, 207

Ca-montmorillonite, 25, 28, 33
Carbonates, 8, 9, 14, 28, 45–48, 119, 123, 126, 127, 130, 143, 147, 151, 185, 187, 195, 207–209
Carbon cycle, 143–151
Carbon dioxide (CO_2), 8, 11–14, 28–30, 58, 60, 81, 123, 145, 148–150, 155, 158, 159, 163, 173–174, 183, 208, 210
Carbon flux, 158
^{14}C data, 25
$CdCO_3$, 193
Chalcedony, 21, 22, 33, 56, 91, 126
Chalcopyrite, 38, 40, 64, 66, 69, 191
Chemical composition of ground water, 25, 32, 90
Chemical equilibrium, 1–49, 118–121
Chemical potential, 4, 5, 83
Chemical reactions, 8, 155
Chimney, 66, 97–99, 103, 105, 110–117, 133
Chlorite, 16, 23, 53, 55, 96, 106, 119, 136, 209
Chloro-complexes, 40, 42, 44, 105
Clay minerals, 121, 133, 186, 193, 207
Cl⁻ concentration, 20, 21, 127, 133
Climate change, 205
Clinozoisite, 107, 108
Closed system experiment, 78
Concentration, 1, 3, 8, 14, 17, 18, 20–22, 25, 28, 29, 32, 33, 35, 37, 39–41, 46, 48, 49, 55, 56, 59, 60, 62, 63, 65, 67, 68, 73, 74, 76, 78–80, 83, 88–91, 93–95, 99, 105, 109, 110, 112–116, 118, 120, 123–128, 131–133, 136, 137, 141, 142, 148, 149, 155, 157–159, 163, 165–167, 174, 176, 181, 184, 186, 187, 189, 191, 193, 197–200, 203, 204
Continental crust, 141
Core, 141, 159
CO_2 supercritical fluid (CSF), 208
Coupled geochemical cycle, 154–157
Coupled models, 87–99
Coupling mechanisms, 73
Cracks, 21, 25, 53, 74, 103, 205
Crust, 17, 53, 105, 108, 118, 122, 133–134, 141, 143, 147, 157, 159, 163, 166–168
Crystal, 15, 16, 44–49, 60, 74, 78, 80, 82, 83, 108, 113, 116, 118, 131, 207, 210
   growth, 113
   ionic radius, 47
   structures, 15, 17
CuO, 8, 10

D

Darcy's law, 84–86, 115
Davies equation, 3
Debye–Hückel theory, 3
Degassing, 59, 149, 158–160, 163, 174
Degassing flux, 163
Degree of supersaturation, 47, 49, 75, 76, 78, 79, 81, 82, 113, 116, 117
Dendritic, 113
Density, 74, 84, 85, 98, 113, 115, 204
Diagenesis, 106, 131
Diaspore, 24
Diatom, 131
Diffusion, 2, 45, 46, 73, 75, 76, 78, 79, 83, 84, 87, 95–99, 105, 113–116, 132, 185, 194, 196, 200, 202, 204, 205, 208
Diffusion coefficient, 75, 76, 84, 115, 204
Diffusion-controlled-mechanism, 75
Diffusion-flow model, 96–99
Diffusion-fluid flow model, 87
Discharge zone, 18, 56, 103, 108
Dislocation, 74
Dissolution, 8, 11, 12, 15, 23, 25, 28, 31, 33, 64, 73–78, 81, 89–91, 93–95, 105, 109, 116, 117, 124, 126, 127, 131, 147, 183, 184, 187, 191, 197, 202, 203, 205, 208–210
Dissolution mechanism, 73–78
Dissolution-recrystallization model, 116–117
Doener–Hoskins distribution coefficient, 48
Doener–Hoskins rule, 49
Dolomite, 123, 147, 148, 195
Double chains, 17
$\delta^{18}O$, 57, 58, 61, 133, 134, 149

E

Earth system, 141
Eh, 36, 189, 190
Elastic model, 46
Electrum, 67, 68
Environmental problems, 173
Epidote, 23, 53, 55, 107
Epithermal gold deposits, 67
Epithermal-type, 36
Equilibrium constants, 8–10, 13, 33–37, 45, 47, 55, 67, 76, 82, 119, 121, 122, 192
Evaporation, 118, 122, 123, 130, 131, 178, 185, 193, 198
Extended Deye-Hückel equation, 18

Index

F
Far-field geologic environments, 202
Feldspar, 4, 8, 9, 11, 17, 19, 20, 22–24, 26–29, 31, 33, 55–58, 90, 91, 95, 105–108, 126, 128, 195, 208, 209
Fick's law, 83
Filling temperature, 61
Flow rate, 18, 25, 115
Flow-through experiment, 78
Fluid flow, 66, 86–95, 99, 110–116, 124, 126, 194, 196
Fluid inclusions, 60, 66, 108
Formation of evaporite, 130
Forsterite, 96, 97
Fossil fuels, 173, 174
Framboidal pyrite, 131
Framework silicates, 17

G
Gabbro, 108
Garnet, 49
Gaseous fugacity, 1
Geochemical cycle, 141–169, 179
Geochemical cycle of minor elements, 165–167
Geochemical modeling, 154, 173, 194, 210
Geosphere, 141, 143, 147, 176
Geothermal reservoir, 18
Geothermal system, 18
Geothermal water, 5–7, 18–23, 59, 60, 79
Gibbs free energy, 46, 47, 77, 99
Gibbs-Kelvin equation, 116
Global carbon cycle, 149, 158–163
The global geochemical cycle, 133, 157
Global sulfur cycle, 163, 165
Gold, 11, 13, 15, 67, 88, 201
Gold-thio-complex, 67
Grain size, 83, 85, 114, 116, 117, 210
Granite, 127, 169, 195, 207
Granitic rock areas, 32, 33
Granitic rocks, 207
Ground water
   chemistry, 29, 32
   scenario, 202–205
   system, 73, 182–193
Gypsum, 61, 64, 118, 119, 128, 151, 157, 187, 188, 207

H
Halite, 128
Halloysite, 25
Hard and soft, acids and bases (HSAB) principle, 40, 42
Hardness and softness of acids-bases, 40
Heat source, 18, 103
^{3}He/^{4}He, 135
Hematite, 11, 38, 64, 69
Homogenization temperature, 61
Hot spot, 163
Humans, 141, 173–210
Hydraulic gradient, 85
Hydrolysis, 15, 34
Hydrosphere, 127, 141, 147, 176, 182
Hydrothermal, 18, 19, 23–25, 36–38, 40–42, 46, 49, 53–56, 58, 59, 61, 62, 64–70, 73, 96, 99, 103–114, 116–118, 130, 132, 134–137, 149, 153, 154, 157–159, 165–169, 207
   activity, 66, 136, 137
   alteration, 18, 19, 23–25, 53, 55–56, 136, 154
   flux, 135, 136
   solution, 18, 19, 38, 55, 56, 58, 59, 64–66, 68, 96, 98, 103–108, 110, 113, 115, 130, 134–137, 154, 157–159, 163, 166–168
   system, 67, 99, 103–118
Hydroxides, 8, 9, 137, 191

I
Igneous rocks, 123, 136, 151
Interstitial water, 132–133
Ion exchange, 11, 15, 47, 73, 79, 89, 121, 122, 186, 195, 198, 205, 207, 208
Ionic activity product, 39
Ionic radii, 3, 42, 43, 47, 48
Ionic strength, 3, 35, 37, 38, 47, 67
Iron
   hydroxides, 193
   oxidizing bacteria, 189, 191
   sulfide, 131

K
Kaolinite, 8, 9, 11, 24–29, 33, 37, 78, 95, 119, 126, 195
Keiko (siliceous ore) zone, 64
K-feldspar, 4, 11, 19–23, 31, 55, 56, 58, 90
Kinetic equation, 74
Kinetics, 66, 73–83, 91, 110–113, 118, 124, 126, 205
K-mica, 4, 11, 19, 21, 37, 56, 58
K-smectite, 23
K&S model, 149, 150

Kuroko deposits, 61, 63, 64, 66, 113, 116, 168
Kuroko ore, 110, 116, 118, 168, 169
Kuroko ore body, 62, 118
Kuroko (black ore) zone, 64

L
Ligand field stabilization energy (LFSE), 40, 42, 43
Light rare earth elements (LREE), 207
Limestone, 123, 127, 195
Lithosphere, 141
Logistic model, 95

M
Magnesite, 148
Mantle, 16, 141, 150, 154, 157–160, 163, 165–167
Marine snow, 130
Mass action law, 18, 35
Mass of water (M), 74, 210
Mass transfer, 73, 74, 83, 87, 95, 99, 103, 141, 173, 182, 205
Mass transfer mechanism, 73–99
Matrix diffusion, 202
Metamorphism, 53
Metastable phase, 5, 82–83, 117–118
Meteoric water, 18, 55
Mg-chlorite, 19, 23, 54, 56, 105, 106, 128, 134
Mg flux, 136
Mg-olivine, 31
Mg problem, 136
Mg-smectite, 134
Microorganisms, 187
Mid-oceanic ridge, 53, 65, 96, 99, 107, 108, 113, 117, 122, 132, 134–137, 148, 149, 157–159, 163, 165, 166, 168
Mid-oceanic ridge basalt, 53
Mineralization, 66, 113
Minerals, 4, 8–17, 19, 21, 23–25, 31, 32, 36, 38, 40, 44, 45, 47, 49, 53, 54, 56–64, 66, 68, 73, 75, 76, 78, 83–86, 97, 99, 103, 106–110, 116–119, 121, 122, 128–134, 136, 147, 148, 168, 174, 185–187, 196, 197, 199, 202, 205, 207, 210
  surface area(A), 74
  surfaces, 15, 78, 191
Mixing, 21, 28, 32, 38, 57–59, 61–70, 87–91, 93–95, 105, 110, 112–116, 118, 124, 133, 142, 196, 199
Molecular diffusion rate, 185
Mole fraction, 1, 44, 47, 62, 63, 121, 192
Morphology, 49, 113, 116

N
N_2, 2
Na·Ca-feldspar, 27
Na-K-Ca geothermometer, 21, 23
Na-montmorillonite, 4, 11, 119, 121, 128
Natural system, 1, 17

O
Oceanic crust, 141, 154
Oko (yellow ore) zone, 64
Olivine, 15, 17, 31, 32, 96, 208
One dimensional vertical model, 200
Opal, 21, 56
Ore constituent elements, 168–169
Ore-forming elements, 105
Ore minerals, 39
Ore zoning, 38
Organic carbon burial, 151
Organisms, 130, 145, 176, 189, 193, 196, 197, 201
Orthosilicates, 17
Ostwald ripening, 82, 116
Otavite, 193
Overpack, 202, 204
Oxidation, 32, 36, 68, 157, 178, 186, 187, 189, 195
Oxidation of pyrite, 186
Oxidation-reduction potential, 189
Oxides, 4, 8, 9, 189, 191, 197, 199
Oxygen (O_2), 4, 5, 8, 9, 11–14, 16, 17, 21, 24, 25, 27–31, 33, 35–38, 40, 57, 68, 69, 77, 81, 89, 91, 96, 97, 99, 107, 108, 110, 112, 117, 119, 120, 123, 126, 128, 130, 132, 144–152, 155, 158–160, 163, 173, 174, 176, 181–183, 186, 187, 189, 195, 196, 207–210
  isotopic exchange equilibrium, 57
  isotopic fractionation factor, 57
  isotopic variations, 57–58

P
Partitioning of elements, 44–49
PCO_2, 30, 81, 123, 150, 183
P cycle, 176, 179
Peclet number, 204
Perfectly mixing non-steady state Model, 196–200
pH, 8, 11, 12, 14, 15, 21, 25, 28–33, 35–38, 40, 41, 47, 58, 65, 67, 68, 75, 77, 80–82, 89, 128, 133, 145, 148, 183, 184, 186, 187, 189–192, 194–196, 208
pH of Lake Water, 192, 194–196

Phosphorus (P), 176
Photosynthesis, 144, 145, 173
Piston flow model, 87, 112
Pitzer's equation, 3
Pitzer's formulation, 3
Plume, 113, 114, 137, 150, 159, 163, 165
Pollution, 122, 173
   of lake water, 194
   in ocean, 200–201
   of river water, 193–194
Polynuclear growth, 79
Pore, 83–86, 91, 204, 208
Pore water, 84, 204
Porosity, 84–86, 95, 98, 115, 125, 128, 204
Precipitation, 8, 15, 28, 33, 39, 46, 48, 49, 54, 56, 59, 66, 68, 70, 73–83, 91, 95, 99, 103, 105, 108, 110–113, 115–117, 130, 136, 147, 148, 168, 173, 193, 197, 199, 202, 205, 207–210
Precipitation-dispersion model, 113–114
Progress variable, 53, 54
Propylitic, 23, 24
Propylitic alteration, 23
Pyrite, 23, 37, 38, 40, 41, 64, 69, 82, 114, 131, 133, 151, 154, 157, 160, 165, 166, 187
Pyrophyllite, 24

Q

Quartz, 4, 11, 15, 17, 19, 21, 22, 24, 33, 56, 58, 64, 74, 75, 83, 90, 96, 97, 107, 108, 110–112, 115, 117, 126

R

Radioactive elements, 202–205, 207
Radiolaria, 128, 131
Rainwater, 25, 29, 31, 88, 123, 177, 180, 182–185, 187, 193, 196
Rainwater-soil reaction, 185–189
Rare earth elements (REE), 116, 207
Rate constant, 15, 74, 76–78, 80, 81, 88, 90, 91, 99, 109, 113, 118, 119, 124, 127, 210
Rate-determining mechanism, 75
Rate-determining step, 74, 75
Rate equation, 76, 77
Rayleigh fractionation, 48–49
Reaction, 3, 4, 6–13, 15, 18, 20, 21, 25–29, 31–34, 36, 40, 42, 44–47, 53, 54, 57–58, 63, 67, 68, 73–82, 84, 87–96, 99, 103, 106, 107, 113, 118–123, 126, 132, 136, 144, 145, 147, 148, 152, 155, 159, 160, 165, 179, 181–187, 189, 192, 193, 195, 196, 205, 207–210

Reaction-diffusion model, 87, 95
Recharge zone, 18, 56, 103, 105–107
Recrystallization, 49, 116–117
Reduction, 15, 32–44, 105, 133, 152, 156, 157, 174, 195, 196
Regular solution model, 1, 1923
Repository site, 202, 204
Reservoir, 18, 21, 103, 105, 107–108, 125, 141–144, 146–148, 153–159, 176, 177, 180
Residence time, 25, 27, 90, 91, 93, 113, 119, 141, 142, 152, 159, 197, 199
Residual gas trapping, 208
Retardation, 203, 204, 207
Reynolds number, 85
Ridge axis, 53, 107, 136
Ridge flank, 136
Ring structure, 17
Riverine fluxes, 136, 165, 194
Rocks, 11, 16, 18, 21, 23–25, 27, 29, 30, 32, 38, 49, 53, 55–58, 62, 83–88, 91, 95, 103, 105, 106, 108, 123, 127, 135, 136, 148, 151, 160, 167, 168, 174, 186, 189, 195, 196, 202, 205–208
Runoff, 124–128

S

Salinity, 3, 38–40, 108
Sandstone, 127, 210
Seafloor hydrothermal system, 103
Seafloor spreading, 148, 149
Seawater, 18, 19, 26, 28, 49, 53–55, 61–66, 73, 103–106, 108–110, 113, 114, 116–119, 121–137, 148, 149, 151, 154–157, 165–168, 194
Seawater-basalt interaction, 134
Seawater-basalt reaction, 54
Seawater chemistry, 118–120, 136
Seawater-rock interaction, 63, 105, 135, 136
Sedimentation, 122, 133, 136
Seepage, 133
Sekko ore body, 61
Sericitic alteration, 23
Serpentine, 96
Sheet silicates, 17
Silica geothermometer, 21
Silicate-carbonate model, 147, 163
Silicates, 4, 8, 11, 16, 17, 28, 74, 76, 78, 88, 90, 105, 106, 119, 123, 127, 147, 151, 187, 189, 208
Single chains, 17
SiO_4 tetrahedron, 16
Smectite, 23, 28, 53, 54, 56, 105, 106, 136, 209

Soils, 153, 182, 183, 187, 191, 207
Solid earth, 160
Solid solution, 1, 18, 20, 44–48, 67, 107, 120, 191, 193, 207
Solubility model, 203
Solubility product, 39, 45–47, 62, 63, 81, 82, 130
Solubility trapping, 208
Spent fuel, 201
Sphalerite, 36
Spiral growth, 79, 82
$^{87}Sr/^{86}Sr$, 63, 64, 106, 108, 109, 116, 159, 160
Steady state, 87, 89, 90, 93–95, 99, 109, 118–119, 125, 132, 142, 143, 145, 146, 155, 159, 185
Steady state model, 196
Stokes equation, 113
Subduction, 148, 149, 158–160, 163, 165, 167
 flux, 163
 zone, 149, 150
Subsystem, 105, 141
Sulfate minerals, 66
Sulfate-reducing bacteria, 131, 152
Sulfides, 8, 36, 39, 40, 66, 96, 110, 116, 117, 131, 151
Sulfur (S), 174, 176–178
Sulfur-carbon-oxygen (S-C-O) cycle, 154–157
Sulfuric acid solution, 24
Sulfur isotope, 105
Sulfur oxidation bacteria, 186
Sulfur-oxidizing bacteria, 132, 152
Superplume activity, 159
Surface energy, 83
Surface free energy, 116
Surface reaction, 78
Surface water, 18
System analysis, 103–137

T
Talc, 96, 97
Thermodynamics, 31

Tortuosity, 84, 115
Total concentration of dissolved sulfur species, 18
Transition state, 74

U
Uranium mine, 25, 207

V
Vein-type deposits, 36–38, 42
Vent, 79, 98, 103, 107, 113–115, 132, 134–136
Viscosity, 84, 85, 113
Volcanic gas, 25, 68, 122, 153, 158, 159, 163, 165–167, 196
Volume diffusion, 83
Volume flow, 85, 88, 93, 113, 124

W
Waste glass, 202, 204, 207
Waste water, 193
Water flux, 135, 136
Water-rock interactions, 23, 173
Water/rock ratio (W/R), 55, 57
Water-rock system, 73, 83, 205
Weathering, 11, 15–17, 26, 28, 88, 90, 92, 95, 122–124, 128, 133, 147, 149, 151, 160, 166, 173, 174, 177, 194, 195, 207
Weathering flux, 156
Wurtzite, 110, 117

Y
Yang module, 47

Z
Zeolites, 56, 209
ZnO, 8, 10

MIX
Papier aus verantwortungsvollen Quellen
Paper from responsible sources
FSC® C105338

If you have any concerns about our products,
you can contact us on
ProductSafety@springernature.com

In case Publisher is established outside the EU,
the EU authorized representative is:
**Springer Nature Customer Service Center GmbH
Europaplatz 3, 69115 Heidelberg, Germany**

Printed by Libri Plureos GmbH
in Hamburg, Germany